Morell Mackenzie

The Use of the Laryngoscope in Diseases of the Throat

Morell Mackenzie

The Use of the Laryngoscope in Diseases of the Throat

ISBN/EAN: 9783337375515

Printed in Europe, USA, Canada, Australia, Japan

Cover: Foto ©berggeist007 / pixelio.de

More available books at **www.hansebooks.com**

THE USE OF

THE LARYNGOSCOPE

IN

DISEASES OF THE THROAT.

WITH

AN APPENDIX ON RHINOSCOPY.

BY

MORELL MACKENZIE, M.D., Lond., M.R.C.P.

PHYSICIAN TO THE DISPENSARY FOR DISEASES OF THE THROAT;
CORRESPONDING MEMBER OF THE IMPERIAL-ROYAL SOCIETY OF PHYSICIANS OF VIENNA,
ETC. ETC. ETC.

"Were it not better for a man in a fair room to set up one great light, than to
go about with a rushlight into every dark corner?"—BACON.

PHILADELPHIA:

LINDSAY AND BLAKISTON.

1865.

ILLUSTRATIONS.

CONTENTS.

INDEX.

THE LARYNGOSCOPE.

CHAPTER I.

HISTORY OF THE INVENTION OF THE LARYNGOSCOPE.

"Honor belongs to the first suggestion of a discovery, if that
suggestion was the means of setting some one to work to
verify it; but the world must ever look upon this last operation as the crowning exploit."—BAIN.

IT may seem strange to some that it was not
till the middle of the last century that an
instrument was invented for examining the lower
part of the throat during life, nor till more than
a hundred years later that that instrument was
sufficiently improved and simplified to be capable
of general application. The dentist's mirror seems
to have been used from time immemorial,* and
polished tubes for passing into the external canals

---

* In the Augustan age, dental surgery had attained a degree
of perfection which implies the employment of mirrors for examining the inner surface of the teeth.—Celsus, lib. vii, cap. xii.

of the body, and thus obtaining an inspection, are of very ancient origin.*

A mere transfer of the dentist's mirror from the mouth to the back of the throat was not however sufficient to give birth to the laryngoscope; and the speculum (which is simply a rigid tube meant to press back the flaccid walls of a straight canal, and thus allow luminous rays to pass through it) was not applicable to the examination of a part situated at an angle to the line of vision. It was only by a combination of these two elements (reflection and illumination) that the interior of the larynx could be seen in the living subject. This fact, together with the circumstance that it was not till comparatively recently, that physicians attempted to discriminate between diseases of the fauces and those of the windpipe, will perhaps account for the non-invention of the laryngoscope at an earlier date. Whatever the cause may be, however, there is no trace of a laryngoscope before the middle of the eighteenth century.

* Some of my readers who have been in Italy may have seen the speculum found in excavating the buried city of Pompeii.

In the year 1743, and probably some years previously, M. Levret, a distinguished French accoucheur, whose highly inventive genius had led him to contrive surgical instruments of almost every description, occupied himself in discovering means, whereby polypoid growths in the nostrils, throat, ears, and other parts, could be tied by ligatures.* It is unnecessary to describe here, the various ingenious instruments which he invented for the purpose, and it is only requisite to observe that in using them he employed a speculum which differed from the various *specula*

---

* "Mercure de France," 1743, p. 2434. The extract from the "Mercure de France," which relates to the employment of the speculum, forms the first article of the appendix to M. Levret's well-known work "L'Art des Accouchemens" (second edition, Paris, 1761). In this article the term "Gozier" is used in one place, and "Gosier" in another. In the latter, the expression used is, "mais pour en appliquer l'usage [of the instrument for carrying the ligature] aux Polypes du Gosier, *situés derrière la voile du Palais*, il a fallu pratiquer . . ." From this, it may seem probable to some, that Levret, in using the term "Gosier," meant the posterior nares. Such an employment of the word would, however, be quite exceptional, and it is much more likely, that he referred to the "throat" generally. In the third edition of Levret's work (the only one I have had the opportunity of consulting) the particular extract from the "Mercure de France," which is quoted above from the second edition, has been omitted. I have to thank Dr. Christie, of Aberdeen, who was the first to call attention to Levret's claims, for very kindly copying the entire extract and placing it at my service.

oris then in use. It consisted mainly of a plate of polished metal (*plaque polie*) which "reflected the luminous rays in the direction of the tumor," and at the same time received the image of the tumor on its reflecting´ surface. It is evident that this little mirror was regarded as a mere appendix of that which Levret considered much more important,—viz. his method of applying ligatures; and that he did not recognize its value as a means of diagnosis in diseases of the larynx. The whole subject was soon lost sight of, and it was not till more than fifty years later that it again excited attention.

Then it was that a certain Dr. Bozzini, of Frankfort-on-the-Maine, made a great sensation throughout Germany, with his invention for illuminating the various canals of the body. About the year 1804, he first made known his ideas, which in the beginning were treated with derision. Gradually, however, the fame of the physician spread, the value of his invention was enormously exaggerated, and not only the professional press, but even political and literary journals, teemed with accounts of the wondrous apparatus. In the year 1807 Dr. Bozzini published a work on the subject of his invention, entitled "The Light-Conductor, or Description of a Simple Apparatus for the Illumination of the Internal Cavities and Spaces in the living Animal Body."* About this time the public seem to have become still more impressed with the value of Dr. Bozzini's invention, and an absurd idea appears to have got abroad that the apparatus would enable practitioners to inspect, not merely the outlets of the body, but even the internal viscera. There was nothing in the work except perhaps its rather ambitious title to encourage this idea; but this did not save it from incurring the wrath of the profession. It is curious that the feeling against the invention should have been strongest in the very city, from which so many of the earliest and most valuable laryngoscopic observations afterward issued. The Faculty of Physicians of Vienna, in concert with the members of the Joseph's Academy, passed a very damaging

---

* "Der Lichtleiter, oder Beschreibung einer einfachen Vorrichtung, und ihrer Anwendung zur Erleuchtung innerer Höhlen, und Zwischenräume des lebenden animalischen Körpers." Von Philipp Bozzini, der Medizin und Chirurgie Doctor, mehrerer gelehrten Gesellschaften Mitgliede, u. s. w. 23 Seiten in Fol. geheftet. Weimar, 1807.

opinion on Dr. Bozzini's invention. They pre-
faced their admonition by remarking that "pre-
mature conclusions were likely to be arrived at
concerning the instrument;" and "that perhaps
even there might be an outlay of money (!!),
which might afterward be regretted." They then
went on to say that "only very small and unimportant parts of the body could be examined;"
that "the illuminated spot was so small—its
diameter being never more than an inch—that
if a person did not know beforehand exactly what
he was to look at, he would not generally be
able to tell what part of the body was presented
to view."*

This was the spirit in which Bozzini's invention was received; a description† of it will show
that it deserved a better fate. It consisted of
two essential parts: 1st, a kind of lantern; and
2dly, a number of hollow metal tubes (specula)
for introducing into the various canals of the

---

* "Salzburg Med.-Chi. Gaz.," Feb. 23, 1807.

† I have not been able to find Bozzini's original pamphlet; but
an abridgment of it appeared in the "Salzburg Medico-Chirurgical
Gazette," February 26, 1807, and another in the seventeenth
volume of Hufeland's "Arzeneikunde." The latter is illustrated
with plates.

body. The lantern was a vase-shaped apparatus made of tin, in the center of which there was a small wax candle. The top of the apparatus was covered; but a large aperture at the upper part, and some holes in its base, allowed sufficient supply of air for the candle; the latter was fixed in a metal tube and forced upward by a spring, after the manner of a Palmer's lamp. In the side of the apparatus there were two round holes, a larger and smaller one, opposite each other. To the smaller one an eye-piece was fixed, to the larger the speculum was fitted. The flame of the candle came just below the level of these two apertures. The mouth of the speculum—a tube of polished tin or silver—was always the same size; but the diameter of the tube beyond varied according to the canal in which it had to be introduced. The apparatus was about thirteen inches high, two inches from before backward, and rather more than three from side to side. These measurements were considered necessary, in order that there should be sufficient space for the candle to burn steadily, and that the lantern should not become too hot. The eye-piece was arranged to fit the eye, so that everything was hidden from view, except the spot seen through the speculum. It may be remarked, that the vase-shaped chamber lined with tin constituted, in fact, two concave mirrors, one behind and the other in front of the reflector; the posterior reflector (if the expression may be used) being perforated by a hole for the eye-piece, and the anterior by another for the speculum.

It is not necessary to enter into details concerning the different canals for which this "simple apparatus" was recommended; but the following quotation* shows that the requisites for making a laryngoscopic examination were fully appreciated by Dr. Bozzini: *"If a person wishes to see round a corner into a part of the throat,†* or behind the palate into the posterior nares, the rays must be broken, and *a mirror is required for illumi-*

---

* Hufeland's " Arzencikunde," Bd. xvii, S. 116.

† The word used is "Schlund." This term is now employed, anatomically speaking, for the pharynx; but it is often used to express the throat generally, and by Hilpert is considered synonymous with "Kehle," the larynx.

nation and reflection." In employing reflected light, Bozzini had the speculum divided by a vertical partition, so that there were, in fact, two canals and two mirrors. One of these mirrors was intended to convey the light, the other to receive the image. We know that this arrangement is quite unnecessary, as one mirror is able to serve both purposes. The annexed wood-cut shows Bozzini's speculum; it is seen to bear a strong resemblance to the instrument invented at a later period by Avery. (*See* Fig. 3.)

FIG. 1.

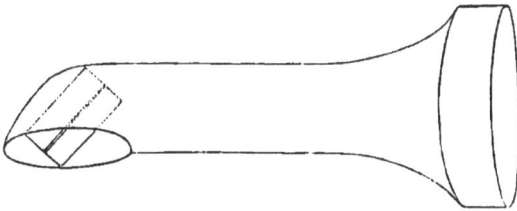

Fig. 1.—BOZZINI'S LARYNGEAL SPECULUM (*after Hufeland*).

The speculum was divided by a vertical partition, and two mirrors were placed at its extremity. In the drawing from which this is taken, the mirrors are directed upward, as they would be, when employed in rhinoscopy.

This ingenious apparatus has ceased to attract attention, and even the historical fact of its former existence was entirely overlooked for several years after the invention of the modern laryngoscope. Whether its neglect was due to the exaggerated expectations, and subsequent disappointment of the public, to the organized opposition of the profession, or to certain constructive defects in the apparatus, it is now impossible to say; but probably they all combined toward the same end. The elements of laryngoscopy were undoubtedly contained in the "Light Conductor;" but it has been justly remarked* by one of the greatest living writers, that "no art is complete, unless another art, that of constructing the tools *and fitting them for the purpose of the art*, is embodied in it." In this instance, the tools were not fitted for the purpose of the art; the latter therefore was never developed, and even the existence of the apparatus soon passed away from the burdened memory of the medical practitioner.

In the year 1827, twenty years after the publication of Bozzini's pamphlet, Dr. Senn, of Geneva, tried to examine the larynx of a little girl, suffering from difficulty of breathing, and

---

* "System of Logic," Introduction, § 7. By John Stuart Mill.

extreme dysphagia. The case was not a favorable one for the trial, and the attempt failed; but as Dr. Senn did not employ any means for throwing a light into the larynx, it is not likely that his efforts would, under any circumstances, have proved more successful. He did not, like Bozzini before and Babington after him, perceive that in laryngoscopy, two factors (illumination and reflection) must always be employed. The following are Dr. Senn's remarks upon the subject: "I had a little mirror constructed for introduction to the back of the pharynx; with it I tried to see the upper part of the larynx,—the glottis; but I gave up its use on account of the small size of the instrument. However, I believe that this method could be employed with advantage in the case of adults, and that in certain cases of laryngeal phthisis, it might assist in diagnosis."* Though this attempt was made in the year 1827, it was not recorded before the end of the year 1829; even then the account of the employment of the mir-

* "Journal des Progrès," 1829, p. 231, note.

ror was not embodied in the text of the report, but was merely appended as a note to the communication. The case was one of considerable interest, both on account of its general features, and especially from its having been one of the first in which a canula had been worn in the trachea for any length of time. It was particularly with reference to this circumstance, that the case had been brought before the Académie des Sciences on the 10th December, 1827.[*] In the published account of the séance there is no mention of any attempt at laryngoscopy.

In the year 1829[†] Dr. Benjamin Guy Babington exhibited, at the Hunterian Society of London, an instrument closely resembling the laryngoscope now in use. Two mirrors were employed by this physician: one, the smaller, for receiving the laryngeal image; the other, larger one, for concentrating the solar rays on the first. The patient sat with his back to the sun, and while the illuminating mirror (a common hand

* "Journal Général de Médecine," tom. cii, January, 1828.
† "Lond. Med. Gaz.," vol. iii, p. 555. London, 1829.

looking-glass) was held with the left hand, the laryngeal mirror—a glass one coated with quick-silver — was introduced with the right. By a very simple mechanism, a tongue-depressor was united with the laryngeal mirror, and thereby one of the most serious obstacles to laryngoscopy was attempted to be overcome. A spring was fixed between the shanks of the laryngeal mirror and spatula, in such a way, that, by pressing the two handles together, the tongue was de-pressed. At a later period (between the years of 1829 and 1835) Dr. Babington abandoned the attractive combination of mirror and spatula, and had mirrors made, which closely resemble those now in general use. The mirrors were made of polished steel, and were, like those now in use, inclined to the shanks at an angle of about 120°. Though Dr. Babington used his laryngo-scope on many patients, there are no cases re-corded in which the instrument was employed.

Priority of publication has long been the estab-lished touchstone, by which the disputed claims of inventors have been tested. Tried by this

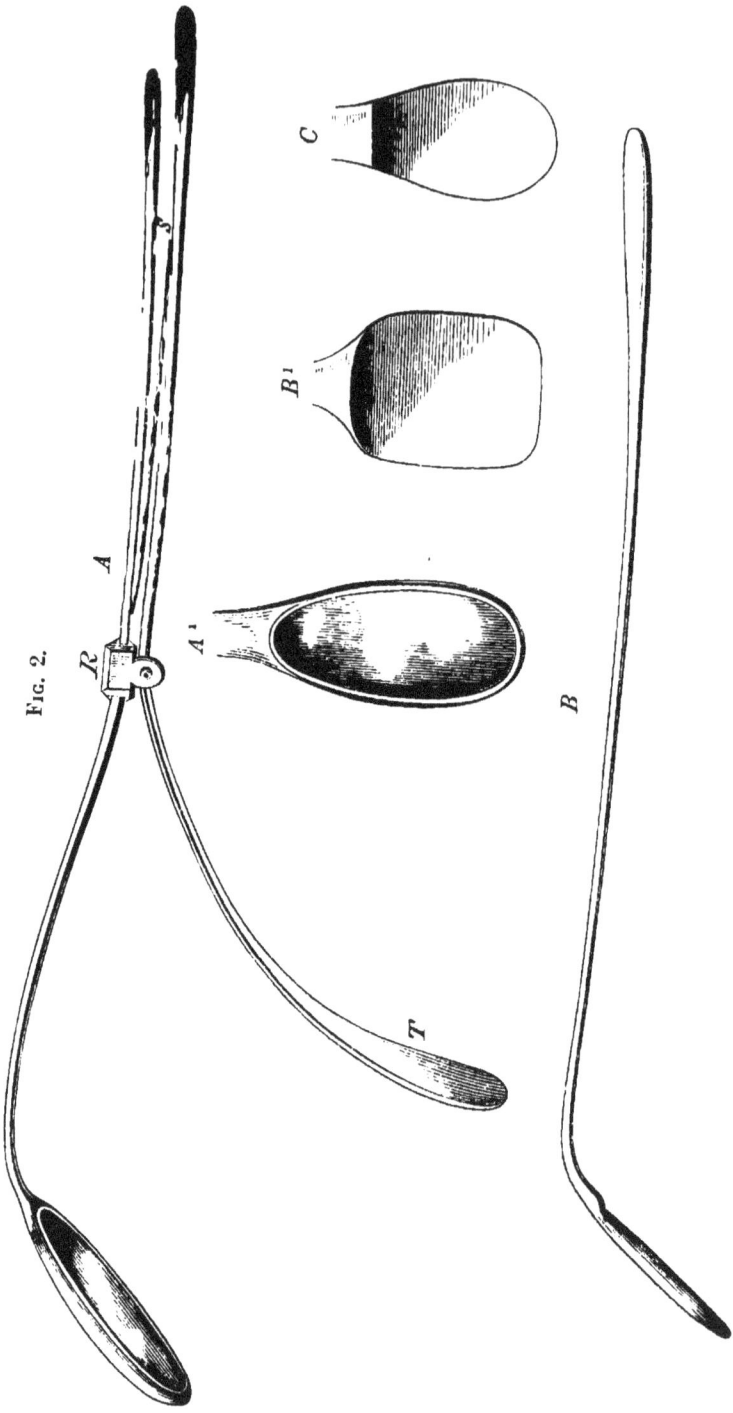

Fig. 2.

DR. BABINGTON'S LARYNGEAL MIRRORS.

A. The instrument from which this drawing was taken was exhibited by Dr. Babington at the Hunterian Society in 1829, and it was again shown, together with Dr. Babington's other mirrors, by me at the Medico-Chirurgical Society, April 27th, 1864.— (*Medical Times and Gazette*, vol. i, 1864, No. 723.)

L. The laryngeal mirror. The steel stem of the mirror widens at one extremity into a frame which contains the glass.

T. The tongue-depressor.

R. A ring which connects the two instruments.

S. A spring which presses the tongue-depressor down, when the two handles are held together.

A¹. Front view of the mirror made in 1829.

B. Side view of steel mirror made between the years 1829 and 1835.

B¹. Front view of the same mirror.

C. Oval mirror made between 1829 and 1835.

(These drawings are the exact shape and size of the instruments themselves.)

criterion. Babington must be regarded as the inventor of the laryngoscope; for while an account of his invention was published in London, in March, 1829, Senn's attempt to examine the larynx was not recorded in Paris till after August in the same year.* The claims of Babington, however, rest on something better than a technical basis; for while Senn merely attempted to employ a *laryngeal mirror*, Babington invented a *laryngoscope*. With the mirror alone it was impossible to see the interior of the larynx; but when a method of illumination was at the same time employed, the inspection became, if not easy, at any rate practicable. The only difference between Dr. Babington's laryngoscope and the one now in general use is, that while in the latter the light is thrown into the larynx (or rather on to the laryngeal mirror) by a circular mirror attached to the head of the operator, in the former the illumination was effected by a mirror held in the operator's hand. Dr. Babington, moreover, does not appear to have

---

* Dr. Senn's letter to the editor of the "Journal des Progrès," which accompanied the report of the case, is dated August. Probably it was not published till a month or two later.

employed artificial light; and his mirrors were of
more clumsy construction than those now used.
Those who have learned to use the laryngoscope,
will readily appreciate the difficulties of illuminating the larynx with a hand mirror; while in this
country, where the sun very often does not shine
brightly for weeks together, the art of laryngoscopy
could never have flourished till artificial light had
been substituted for the uncertain solar rays.

In the year 1832,* while Babington was still
working with his "glottiscope," to use the term employed by him at the time, Dr. Bennati, of Paris,
asserted his ability to see the vocal cords. A
mechanic named Selligue, who was suffering from
laryngeal phthisis, had invented "a double-tubed
speculum, of which one tube served to carry the
light to the glottis, and the other to bring back to
the eye the image of the glottis reflected in the
mirror, placed at the guttural extremity of the

---

* "Recherches sur le Méchanisme de la Voix Humaine," p. 37,
note. Bennati uses the expression "au moyen d'un speculum que
j'ai imaginé." As Trousseau, however, speaks emphatically of *Bennati's experiments with Selligue's mirror*, and as there is no doubt
that Selligue did invent a laryngeal mirror, I have adopted the view
that his instrument was employed by Bennati.

3

instrument." A complete recovery rewarded the ingenious patient for his clever invention, and with this instrument Bennati professed to be able to see the glottis. Trousseau, however, disbelieved his statements, and devoted several pages of his well-known work[*] to prove, that the epiglottis formed an insuperable impediment to a view of the interior of the larynx. This renowned physician had an instrument constructed after the model of Selligue's, but he does not appear to have attempted its use. It is worthy of note, that Selligue's laryngeal speculum closely resembled that of Bozzini, for while the latter was made in one tube divided by a vertical partition, the former consisted of two tubes.

In the year 1838,[†] M. Baumés exhibited at the Medical Society of Lyons, a mirror about the size of a two-franc piece, which he described as being very useful for examining the posterior nares and larynx.

In the year 1840,[‡] Liston, in treating of œdema-

[*] "Mémoire sur la Phthisie Laryngée." Par MM. Trousseau et Belloc. "Mémoire de l'Académie de Médecine," tome vi, 1837.

[†] "Compte Rendu des Travaux de la Société de Médecine de Lyons, 1836-38," p. 62.

[‡] "Practical Surgery," third edition, p. 417. 1840.

tous tumors which obstruct the larynx, observed as follows : " The existence of this swelling may often be ascertained by a careful examination with the fingers, and a view of the parts may sometimes be obtained by means of a speculum,—such a glass as is used by dentists on a long stalk previously dipped in hot water, introduced with its reflecting surface downward and carried well into the fauces." When the real art of laryngoscopy was founded almost twenty years later, the name of our talented countryman was prominently associated with its invention. But it is obvious from the above passage, that Liston never contemplated an inspection of the vocal cords. It is plain that in his estimation the sense of touch was more to be relied on than that of sight; and the fact that the fingers were to be used, indicates pretty clearly that Liston was referring rather to the epiglottis than the parts below.

In the year 1844,[*] Dr. Warden, of Edinburgh,

* Royal Scottish Society of Arts. Description, with illustrations, of a Totally Reflecting Prism for illuminating the open cavities of the body, &c., &c. May, 1844. See also "Lond. Med. Gaz.," vol. xxxiv, p. 256.

conceived the idea of employing a prism of flint glass for obtaining a view of the larynx. The success which had attended his efforts to inspect the membrane of the tympanum, induced him to apply the principle of the prism to other canals. He reported two cases* in which he considered that he had had "satisfactory ocular inspection of diseases affecting the glottis." The possibility of inspecting the larynx in this way admits of no doubt,† but Dr. Warden's method of employing the prism was not calculated to bring about very favorable results. The particulars of one of the cases referred to are given, but in the other "the appearances were so far similar as to render their detail unimportant." The patient whose case is narrated was a lady, "who had been the subject of medical treatment for chronic inflammation of the pharynx of nearly a year's duration;" the inflammation had latterly spread in the direction of the glottis, and painful deglutition and paroxysms of suffocation now supervened. "After the pre-

---

* "Month. Journ. Med. Science," July, 1845, p. 552.
† See page 44.

liminary examination and *quietening the irritability of the parts by touch with the finger*, there was no longer any impediment or inconvenience experienced from the tendency to retching. . . . .

. . . . . . *The dilator faucium was employed to depress the tongue and expand the isthmus of the fauces.*" The result of the examination was that the epiglottis was seen to be very much thickened and inflamed, "but it was only when efforts to swallow were made or repeated that the arytænoid cartilages, in a similar condition of thickening, were raised out of concealment, and brought brilliantly to show their picture in the reflecting face of the mirror." For the purpose of illumination Dr. Warden employed "a powerful argand-lamp, with a large prism attached, so as to throw the full light of the lamp into the fauces and pharynx." That is to say, instead of the two plane mirrors we use (one for illumination, and the other for reflection), he employed two prisms. In concluding the report of these cases Dr. Warden remarks that "the experience afforded by both gives ground for the same conclusion, that the instrument made use of can

have no farther range than the bottom of the pharynx and mouth of the glottis,* and of the latter
only so often as it is raised from its natural depth,
by the contraction of the muscles employed in the
act of deglutition. By this means, therefore, we can
obtain no assistance in the investigation or treatment of diseases below the pharynx." It is not
surprising that Dr. Warden should have expressed
himself thus unfavorably concerning his attempts
to examine the larynx. What with "quietening
the irritability of the throat by touch with the
finger, depressing the tongue, dilating the fauces,
and encouraging the patient to swallow," it was
utterly impossible for him to have succeeded. No
disciple of Czermak could hope to see the vocal
cords were he to prepare his patient in the way
described by Warden; and when we remember how
limited was his experience, and how imperfect his

---

* In using the expression "mouth of the glottis," Dr. Warden
probably meant the upper opening of the larynx, that is to say, the
opening bounded by the ary-epiglottidean folds. He could not possibly have imagined that in deglutition the true vocal cords would
be left uncovered by the epiglottis. It should be borne in mind
(especially in reading cases reported a few years ago), that the term
"glottis," now very properly confined to the aperture bounded by
the true vocal cords, had till quite lately a very vague signification.
—See "Dunglison, Dict. Med. Science."

instruments, the appearances described by him can scarcely be regarded otherwise than the baseless fabric of a very imperfect vision.

In the year 1844, while Dr. Warden was still trying to employ the prism for examining the various canals of the body, Mr. Avery, of London, was seeking to accomplish the same end with the aid of the speculum and reflector. In principle Mr. Avery's laryngoscope was very similar to that now in use; and even in its details it did not differ widely from the modern instrument. Like Bozzini forty years previously, Mr. Avery perceived the value of artificial light, and like Czermak after him he employed a large circular reflector, perforated in the center for concentrating the luminous rays on the laryngeal mirror. The reflector was attached to a frontal-pad; and this was retained in its place by two springs which passed over the operator's head to the occipital protuberance, where there was a counter-pad. There were two defects, however, in Avery's apparatus: the one was, that the laryngeal mirror (instead of being fixed to a slender shank) was placed at the end of a speculum; the other, that instead of employing the reflector for receiving the rays from a lamp placed on the table or elsewhere, Avery used his large circular mirror for the purpose of increasing the luminous power of a candle held near the patient's mouth. This candle was (like Bozzini's) a miniature Palmer's lamp, and was also attached to the frontal-pad.

A piece of bent wire terminating in a circular loop projected from the candle-lamp, and was meant to steady the speculum and keep its axis in a line with the hole in the reflector. The reflector was five inches in diameter, and the apparatus which had to be worn by the operator weighed altogether nearly a pound. For those who preferred it, however, the candle-lamp and reflector could be fixed into the top of the box, which contained the apparatus when not in use. The candle and the reflector, fixed in this way and placed on a table, bore a strong resemblance to the "Light-Conductor" of Bozzini, except that in the latter the light was entirely enclosed within the cavity of the vase-shaped lantern. By an ingenious double-rack movement, the reflector could be made to move either laterally or horizontally. This arrangement allowed for considerable range of distance between the nose and eyes, a high or low forehead, &c., and thus permitted different people to employ the same reflector, and still have the circular hole always opposite the pupil. It was difficult, indeed almost impossible, to introduce Avery's speculum, without irritating the base of the tongue and other contiguous parts, and thus causing a disposition to vomit. This feature alone would have been sufficient to insure the failure of Mr. Avery's attempts at laryngoscopy, had not the cumbersome reflecting apparatus combined to produce the same result. The resemblance which Avery's laryngoscope bears to Bozzini's on the one hand, and to Czermak's on the other, is very striking. In all of them artificial light, circular reflectors, and small laryngeal mirrors were used. In the laryngoscope of Bozzini and Avery, the lamp and the reflector are combined, while in the modern instrument they are separate. The laryngeal mirror of Bozzini and Avery was placed at the end of a speculum; Czermak's was a modification of the dentist's mirror. Mr. Avery's invention was not placed

FIG. 3.

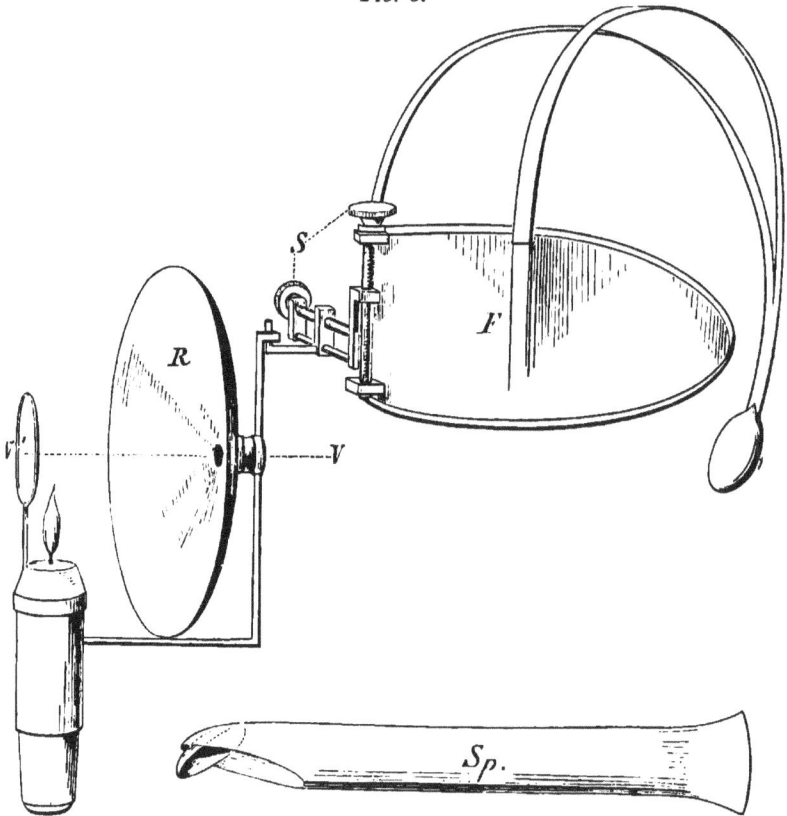

Fig 3.—AVERY'S LARYNGOSCOPE.

F. One side of the frontal-pad which supports the mirror. From it a double spring passes backward to a counter-pad which, when the instrument is worn, rests under the occipital protuberance. In the drawing, the occipital-pad is drawn forward by the unopposed strength of the spring.

S. Screws by which the reflector can be made to move laterally and perpendicularly.

R. Reflector.

VV¹. Line of vision.

Sp. The speculum.

on record* till some time after the modern laryngo-scope had come into use. The laryngoscope from which the drawing on the preceding page was taken was supplied by Messrs. Weiss to the London Hos-pital, in the year 1846.

In the year 1854,† "the idea of employing mir-rors for studying the interior of the larynx during singing" occurred to M. Maunal Garcia. He had often thought of it before, but believing it impracti-cable, had never attempted to realize the idea. M. Garcia, though long a distinguished singing-master in London, was a Frenchman by birth, and a Spaniard by descent; and though his observations with the laryngoscope were first published in England, they were first made in France.

In the month of September, 1854, while Garcia was spending the holidays in Paris, he determined to clear up his doubts concerning the possibility of inspecting the larynx. His efforts were crowned with success, and the following year he presented a

* "Med. Circ.," vol. xx, June, 1862; and Introduction to the Art of Laryngoscopy," by Dr. Yearsley. London, 1862.

† "Notice sur l'Invention du Laryngoscope." Par Paulin Rich-ard. Paris, 1861.—See M. Garcia's letter to Dr. Larrey, dated May 4, 1860. (Page 12 in Pickard's Pamphlet.)

paper to the Royal Society of London, entitled "Physiological Observations on the Human Voice."* This paper contained an admirable account of the action of the vocal cords during inspiration and vocalization; some very important remarks on the production of sound in the larynx, and some valuble reflections on the formation of the chest and falsetto notes. M. Garcia's laryngoscopic investigations were all made on himself; indeed, he was the first person who conceived the idea of an autoscopic examination.

His method, which he believed had never been employed by any one before, consisted in introducing a little mirror, fixed to a long stem, suitably bent, to the top of the pharynx. He directed that the person experimented upon should turn toward the sun, so that the luminous rays falling on the little mirror should be reflected into the larynx;† but he added in a foot note, that "if the observer

---

* "Proc. Royal Soc. London," vol. vii, No. 13, 1855. "Philosoph. Magazine and Journal of Science," vol. x, p. 218, and "Gaz. Hebdom. de Méd. et Chir.," Nov. 16, 1855, No. 46.

† It is worthy of note, that Garcia really never followed this plan, but, in point of fact, always used a second mirror for throwing the solar rays on to the laryngeal mirror. In the mirror which he used as reflector, he also saw the autoscopic image.

experiments on himself, he ought, by means of a second mirror, to receive the rays of the sun, and direct them on the mirror which is placed against the uvula." In practicing auto-laryngoscopy after the manner of Czermak, three mirrors are employed: one for illumination, another for introducing to the fauces, and a third to enable the observer to see the image in the mirror held in his own throat. Garcia employed only two mirrors: a small one at the end of a long stem for introducing to the pharynx, and a large one which served the double purpose of illuminating the little mirror, and enabling the operator to see the image formed on it. It will be seen that Garcia's method was precisely similar to that employed by Babington; the one, however, limited his observations to his own larynx, the other never made an attempt at auto-laryngoscopy. Garcia's communication to the Royal Society, though causing little stir at the time, was destined to experience a fate in many respects similar to that which befell the paper of our countryman, Mr. Cumming.[*] Treated with apathy, if

* "Transactions of Med. Chir. Soc.," 1846.

not with incredulity in England, both   s passed
into the hands of foreign professors, and while
Helmholtz matured the ophthalmoscope, Czermak
developed the laryngoscope.

In the year 1857, during the summer months,
Dr. Türck, of Vienna (who had read Garcia's paper),
endeavored to employ the laryngeal mirror in the
wards of the General Hospital. He was not successful, however, at first, and at the end of the autumn
he seems to have abandoned his fruitless attempts.
Trusting entirely to the solar rays, having no apparatus (no second mirror) for concentrating the light
on the laryngeal mirror, and the latter being a
clumsy hinged instrument, it was scarcely possible
for him to succeed. When at a later period, however, Czermak proved the practical value of the
laryngoscope, Türck put forth his claims to priority.
Nevertheless, in the very communication[*] in which
he asserted his pretensions, he observed that " he
was very far from having any exaggerated hopes
about the employment of the laryngeal mirror in
practical medicine." This unfortunate remark shows

---

[*] "Zeitschrift der Ges. der Aerzte zu Wien.," April 26, 1858.

that he did not even then recognize the value of the laryngoscope.

In the year 1857, in the month of November, Professor Czermak, of Pesth, borrowed from Dr. Türck the little mirrors which that gentleman, in spite of the exhortation of his friends, had thrown aside as useless.[*] In a short time his superior genius, untiring perseverance, and natural dexterity enabled him to overcome all difficulties. When the dentist's mirror passed into the hands of Dr. Czermak, the examination of the larynx was dependent —so to speak—on the clock and the barometer, but he soon relieved it from both these troublesome monitors. Artificial light was substituted for the uncertain rays of the sun; the large ophthalmoscopic mirror of Ruete was used for concentrating the luminous rays; the awkward hinge which united the laryngeal mirror to its stem was dispensed with; and mirrors were made of different sizes. Thus it was that Czermak created the art of laryngoscopy. Others before him had contrived instruments, with

* Professor Brücke's Letter to Czermak. "Selected Monographs: New Syd. Soc.," vol. xi.

which they had sometimes succeeded in inspecting
the interior of the larynx, but "the tools fitted for
the art" of laryngoscopy were not constructed before his time. His first publication appeared in
March, 1858,* and a month later a very important
paper of his was brought before the Academy of
Sciences of Vienna.† In claiming for Czermak the
honor of having so modified the laryngoscope, that
its application became comparatively easy, it would
not be right to withhold from Dr. Türck the merit
of having patiently and productively worked at the
subject at a later period. A careful investigation
of facts and dates, however, must convince every
disinterested person, that Türck's subsequent successful labors were prompted by the proofs which
Czermak had given of the value of the laryngoscope.

Czermak's investigations were at first confined to
his own larynx, and his success must in part be
attributed to his great physical advantages. Pos-
sessed of a most capacious pharynx, small tonsils

---

* "Wien. Medizin. Wochenschrift."

† "Physiolog. Unters mit Garcia's Kehlkopfspiegel," mit iii.
Tafeln. Sitzber d. k. k. Akademie d. Wiss in Wien" vom April, Bd.
xxix, p. 557 (Afterward reprinted in a separate form.)

and uvula, and a large laryngeal aperture, it would be difficult to find a subject better suited for laryngoscopy. Notwithstanding the beautiful simplicity effected by Czermak in the details of the laryngoscope, the · profession might not have become impressed with the value of the instrument, had not his brilliant demonstrations delighted and astonished the medical public throughout Europe. The general employment of the laryngoscope in practical medicine must be attributed not less to his enthusiastic and universal teaching—to his brilliant demonstrations and personal influence, than to his entire remodeling of the instrument itself. The fact that no improvement has been made in the mechanism of the laryngoscope for the last five years, though a great number of practical men in all parts of the world have been constantly working at the subject, is the strongest testimony to the value of Czermak's labors.

CHAPTER II.

DESCRIPTION OF THE LARYNGOSCOPE.

DEFINITION.—An instrument for obtaining a view of the interior of the larynx during life. It consists of two parts: 1st, a small mirror fixed to a long slender shank, which is introduced to the back of the throat; and 2dly, an apparatus for throwing a strong light (solar or artificial) on to the small mirror. For thus projecting the luminous rays, most laryngoscopists employ a second (larger) mirror, which reflects the light from a lamp or the solar rays. When artificial light is employed, this illuminating mirror is slightly concave; when sunlight is used, its surface is plane.

SECTION I.—*The Laryngeal Mirror.*

THE laryngeal mirror may be made of polished steel, or of glass backed with amalgam. Though, on theoretical grounds, the steel mirrors give the more perfect image, they so readily become tarnished and rusty from the least moisture, are so immediately spoiled by the accidental contact of the medicated solutions used in treating laryngeal disease, and so soon become scratched in cleaning, that they are not found convenient in practice. The glass mirror is generally mounted in German silver; for though the metal is too favorable to the rapid cooling of the mirror and the consequent deposit of moisture upon it, it is more easy to fix the shank of the instrument to a frame of metal than to any other substance of inferior conducting power. The mirrors should not be more than one-twentieth of an inch in thickness.

The reflecting surface of the laryngeal mirror may vary from half an inch to an inch and a quarter in diameter. It is well to be provided with at least three mirrors, varying in size between the dimensions specified.

For ordinary purposes a mirror about eight-tenths of an inch in diameter will be found most convenient. It may be of square, circular, or oval shape. The circular mirrors cause least irritation, except when enlarged tonsils are present. In these cases the oval mirrors are most suitable. The use of concave mirrors for magnifying the laryngeal image has been suggested (Wertheim); but while they only enlarge the image very slightly, their employ-

ment is objectionable optically, on account of the
varying distance from the mirror of the parts reflected (Türck). The shank of the mirror should
be of German silver; it ought to be about four
inches in length, and one-tenth of an inch in thickness. It should be soldered* to the back of the
mirror, so that the latter forms with it an angle
of about 120°. The shank or stem of the mirror
slides into a hollow wooden handle, and is fixed
there by a screw. By this arrangement the stem
can be made shorter or longer, according to the
depth of the mouth from before backward in different cases. The handle should be about three inches
in length, and rather more than a quarter of an inch
in thickness.

By my directions some totally reflecting prisms
have been constructed,† with which I have frequently examined patients and demonstrated my

---

* In the construction of the instrument, the shank should be bent
at the desired angle before it is soldered to the back of the mirror;
for if fixed to the mirror first, the angle (instead of being formed at
the junction of the mirror and the shank) is obliged to be about one-
tenth of an inch or more from the mirror. The space (viz. that
between the mirror and the angle of the shank) which is thus lost
would afford room for the employment of a larger mirror.

† By Mr. Ladd, of Beak Street.

own larynx. As the base of the prism is, however, necessarily almost a third larger than its refracting surfaces, the use of this kind of mirror implies a considerable loss of space. The prismatic mirrors are also much more difficult to introduce than the common flat ones, and the inferior surface of the prism is extremely likely to come in contact with the tongue; this, of course, for the time interferes with the passage of light through the prism. In the application of remedies to, or at all events in delicate operations on, the larynx, the projecting triangle of the prism is likely to be in the way. Finally, the prismatic mirrors are much more expensive and more easily damaged than the flat ones. These conclusions from my experiments with prismatic mirrors are briefly related here for the purpose of deterring others from fruitless trials of a similar character.

SECTION II.—*Illumination.*

For throwing a light on to the laryngeal mirror, and thus into the larynx, it will be found most convenient to employ a circular mirror about three inches and a half in diameter, with a small hole in

FIG. 4.

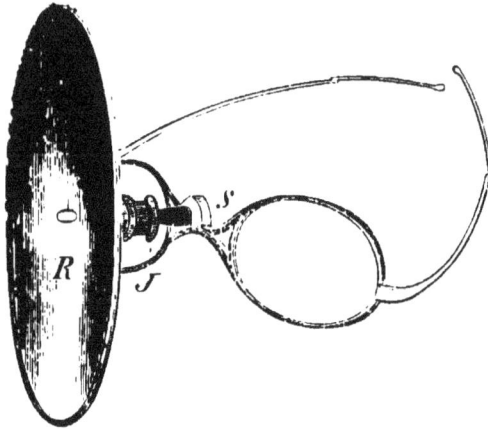

Fig. 4.—REFLECTOR ATTACHED TO SPECTACLE-FRAME.

At the back of the reflector (R) is a small cup, into which a ball connected with the spectacle-frame fits. A ring is screwed over the ball and the joint is thus formed at J.

its center.* When artificial light is employed, the mirror should be slightly concave and have a focal power of about fourteen inches; but when solar light is made use of, the surface of the mirror should be plane. The mirror may be attached in some way

* The reflector should not merely be left unsilvered in the center, but should be actually perforated. In the former case, the glass makes a slight focal inequality between the two eyes. Laryngoscopes, made in every respect according to my directions, are sold by Mr. Krohne, 241 Whitechapel Road.

to the operator's head, or fixed to a horizontal arm
which is connected with the body of the lamp (Tobold). The former plan is by far the most convenient, and the mirror may be worn either opposite
one of the eyes (Czermak), in front of the nose and
mouth (Bruns), or on the forehead (Johnson, Mason,
&c.). Of these positions, the first is on theoretical
grounds the most perfect, the last the easiest in
practice. It is only in the first position, however,
that the observer can look through the hole in the
reflector; if, therefore, either of the other methods
is practiced, the reflector need not be perforated.
The reflector may be attached to the operator's head
either by a spectacle-frame (Semeleder), or by a
frontal band (Kramer). In either case the mirror
should be connected with its support by a ball-and-
socket joint. The hole in the center of the reflector
should be oblong, and when placed in front of the
eye, its long diameter should correspond with the
long diameter of the eye. A hole of this shape
allows for the varying distance between the nose
and eyes in different people, and for the varying

position of the center of the reflector, in its different degrees of inclination.

It should be remarked that, though the employment of the reflector greatly facilitates the inspection of the larynx, a laryngoscopic examination can be effected without it. In this case a strong light must be thrown directly on to the laryngeal mirror.

Any lamp that gives a bright steady light answers the purpose perfectly well. Many of the most valuable observations have been made with a common "moderator." An argand gas-burner will be found very convenient, especially if constructed so that it can be fixed at different heights. The power of the light may be advantageously increased by one or more lenses placed in front of the flame.

Various lamps or lanterns have been recommended by different foreign laryngoscopists (Tobold, Voltolini, Moura-Bourouillou, &c., &c.), but the arrangement of lenses in each of them is only applicable to the particular lamp for which it was contrived. This serious objection to the various kinds of illuminating apparatus hitherto in vogue, led me to contrive a light-concentrator of more extensive application.* It not only gives a very brilliant light, but is at the same time much smaller, and therefore much more

Fig. 5.

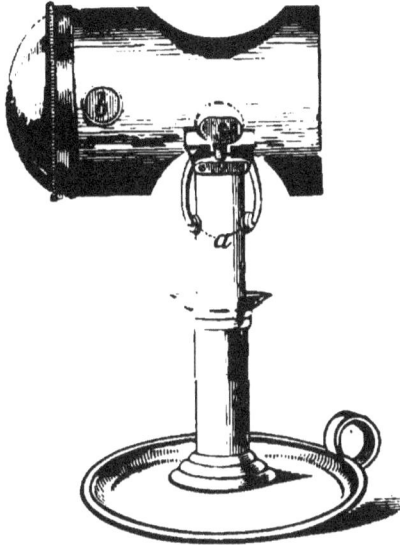

Fig. 5.—The Light-Concentrator.

In the drawing, the concentrator is fixed on to a candle by means of the two arms(*a*). In using a lamp, the arms embrace the chimney.

portable than any of those hitherto in use, and it can be employed with any kind of lamp, or even a candle. It consists of a small metal cylinder three and a half inches long, and two and a half in diameter. This is closed at one end, and at the other

* These "light-concentrators" are made for me by Mr. Mayer, of 51 Great Portland Street.

there is a plano-convex lens, the plane surface of which is next the flame. The lens is two and a half inches in diameter, and is about one-third of a sphere. In the upper and under surfaces of the cylinder (opposite each other) are two round apertures, two inches and a quarter in diameter. These holes are not equidistant from the two ends of the tube, but so near to the closed extremity, that a line passing perpendicularly through their centers would be about two inches and a half from the plane surface of the lens. This is the proper focal distance of the lens, and rays of light pass through in parallel direction. At the lower part of the tube are two semicircular arms, which by means of a screw at the side can be made to grasp tightly the largest lamp-chimney, an ordinary candle, or even the narrow stem of a single gas jet. The practitioner, therefore, who, in visiting patients, carries my light-concentrator, can always feel certain of being able to illuminate the fauces. The apparatus is passed over the chimney, till the center of the lens is opposite the most brilliant part of the flame, and then by a few turns of the screw, the concentrator is fixed in position. When a candle is employed, the flame is in the center of the tube.

In the side of the tube near the lens are two ivory knobs covered with cork, which enable the practitioner to hold the concentrator and remove it from the lamp, even when it is extremely hot. For the consulting-room the light-concentrator may be most advantageously employed either with an argand gas-burner, a parafine, moderator, or reading lamp. The latter kind of lamp made to burn parafine, or having an argand gas-burner, will be found most convenient, as it can be fixed at any height.

The light of a candle, strengthened by this concentrator, will be found to equal that given by an ordinary lamp. When the practitioner has only a center gaselier at his command, the light-concentrator should be applied to the only jet which is lighted, and as it is not generally possible to pull a gaselier sufficiently low down to make the examination in the ordinary way, under these circumstances, both patient and practitioner must stand upright.

Besides the concentrator just described, I have had a smaller illuminating apparatus constructed, which is called my "miniature light-concentrator." The principle is the same in both; but in the latter the metal cylinder is only two inches in length, and an inch and a half in diameter; it is only suited for the small parafine lamp, which is sold with it. This lamp, which measures only four inches from its foot to the top of the chimney, is like a little phial, and has a metal screw stopper, so that it can be carried about with safety. Of the two, I recommend my larger concentrator, because in using the smaller apparatus, the body of light, though very brilliant, has, of course, a very small diameter; and if the practitioner is not very experienced, he will find it rather difficult to keep the reflector in the exact line of light. In the construction of an apparatus for increasing the light, it may be well to observe,—1st. That if a lens is used, it should always be placed at its exact focal distance from the flame: this causes the luminous rays to pass through the lens in parallel direction, and thus throws a body of light a considerable distance. 2dly. That since, when rays fall on a convex surface, a certain number are refracted, and do not pass through it, the plane face of a plano convex lens should be placed next the flame. 3dly. That no mirror or reflector should be placed behind the flame for the purpose of strengthening the light; for any scratches or spots on it are apt to be reflected on to the illuminating mirror, and thence again on to the laryngeal mirror. This, of course, interferes with the distinctness of the laryngeal image.

It has been already observed that the employment of a reflector is not absolutely necessary for throwing a light on to the laryngeal mirror. When the observer does not make use of a reflector, the lamp must be placed very near to the patient's mouth, or else the luminous rays must be projected from a lamp in less close proximity by a lens placed in front of the flame. For this purpose either an ordinary plano-convex lens may be used, or a large glass globe about six inches in diameter, filled with water. The latter kind of concentrator (the so-called Schuster-kugel) was first recommended by Türck, and afterward adopted by Stoerck; but while the former has abandoned its use in favor of the reflector, the latter still employs it. This apparatus is also recommended by Dr. Walker,* of Peterborough, who has greatly improved it by substituting an elegant metal frame for the cumbersome wooden affair of Stoerck. It gives a brilliant light, which is most intense, about twenty inches from the globe. As it is quite impossible to carry this enormous glass globe about, its use is necessarily confined to the practitioner's consulting-room.

A comparison of the respective merits of direct and reflected light, will show that the advantages are principally on the side of the latter. For merely making an examination, it is quite possible to employ a powerful direct light, but when both hands are employed, as in the application of remedies to the larynx, the management of the light becomes much more troublesome. When the observer has the reflector opposite his eye, it is obvious that the visual and luminous rays pass in precisely the same direction, so that when the larynx is illuminated, the observer can see, and when the mirror is inclined at the proper angle for seeing, the larynx is also illu-

---

* "The Laryngoscope and its Clinical Application." By Thomas J. Walker, M.D., Lond., &c. London, T. Richards, p. 13.

minated. When, however, the reflector is placed opposite the forehead or nose, the lines of vision and illumination, though very near each other, do not correspond precisely; when the reflector is thus employed, therefore, the larynx may be illuminated, but the observer may not be able to see. Nevertheless the angle of inclination between these two lines (viz. that passing from the forehead or nose, and that from the eye) is so very slight, that practically it is not a matter of much importance which method is employed. When, however, the luminous ray, instead of corresponding, or nearly corresponding with the visual ray, forms a considerable angle with it—as it must do when direct light is employed— there is a great probability of the two rays not falling within the area of the larynx. Again, in employing direct light, as the rays must pass to the mirror from the side (instead of from before backward, as in using reflected light), they have also a lateral deflexion, and are thus likely to illuminate only one side of the larynx. In employing direct light, the side of the cheek often throws a shadow on the laryngeal mirror, and in the application of remedies the practitioner is apt to get in his own light. Practice may overcome all these difficulties, as has been proved by Dr. Walker; but the creation of unnecessary obstacles cannot be recommended.

The solar rays, or diffused light, on a bright day may be concentrated on the laryngeal mirror. In the former case the surface of the reflector must be plane, in the latter the usual concave mirror may be used. The patient should sit with his back turned obliquely to the window, and the practitioner oppo- site him. The sunlight in this way passes over the patient's shoulder to the reflector, and is thence pro- jected on to the laryngeal mirror. In other respects the examination is conducted in the same way, as when artificial light is used.

Before finally dismissing the subject of illumina- tion, a few remarks may be made on what has been called illumination by transparency.

If sunlight is concentrated on the side of the neck, and the laryngeal mirror is then introduced, a more or less distinct image is obtained. Even under most favorable circumstances, however, where the neck is thin and long, the image is not sufficiently clear to be of any real value; while if the neck is short and muscular, or the glands are at all enlarged, nothing at all can be seen. This kind of illumination was first suggested by Czermak, though he does not attach any importance to it.

5

CHAPTER III.

THE ART OF LARYNGOSCOPY.

The proper employment of the already described instruments constitutes the art of laryngoscopy. It may be practiced by the physiologist for investigating the healthy appearance and normal action of the larynx, or by the physician for inspecting, and if possible improving the condition of the parts when diseased. It is only with reference to the last purpose that it will be here considered.

SECTION I.—*Principles of the Art.*

THE only principle concerned in the art of laryngoscopy is the optical law, that when rays of light fall on a plane surface, the angle of reflection is equal to the angle of incidence. A small mirror is placed at the back of the throat, at such an inclination that luminous rays falling on it are projected into the cavity of the larynx; at the same time the image of the interior of the larynx (lighted up by the luminous rays) is formed on the mirror, and seen by the observer. The mirror is held obliquely, so that it forms an angle of rather more than 45° with the

horizon. The plane of the laryngeal aperture (bounded by the epiglottis, the ary-epiglottidean folds, and
the arytænoid cartilages) is also oblique, the epiglottis being higher than the apex of the arytænoid
cartilage.

The annexed diagram shows the position of the
different parts, and explains their reflection. Let *m*

FIG. 6.

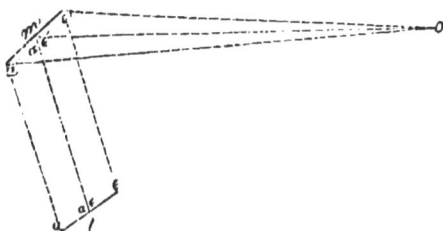

Fig. 6.—Diagram showing the relative positions of the planes of the
larynx and laryngeal aperture.

represent the plane of the laryngeal mirror, *l* the
plane of the upper opening larynx, and *o* the observer. In the plane of the larynx, *a* represents the
arytænoid cartilages, *ae* the ary-epiglottidean folds,
and *e* the epiglottis; the rays from these parts im-
pinge on the mirror, as *á*, *aé*, and *é*, and are thence
reflected to the observer at *o*. Thus the epiglottis,
which is really the highest in the throat, appears at

the upper part of the mirror, the ary-epiglottidean appear rather lower and at each side of the mirror, while at the lowest part of the mirror are the arytænoid cartilages. These remarks apply to the antero-posterior reflection.

The lateral relation of parts in the image must now be considered. The mirror being placed above and behind the laryngeal aperture, the rays of light proceeding from the larynx pass directly upward and backward, and the patient's right vocal cord is seen on the left side of the mirror, and the left vocal cord on the right side of the mirror (just as the patient's right hand is opposite the observer's left, and his left hand opposite the observer's right). In examining a laryngoscopic drawing, a person must not make his own larynx the mental standard of comparison as regards right and left, but must recollect that the picture represents an image formed on a mirror held obliquely above and rather behind the larynx of another person. When the observer is looking at the mirror, he is not likely to make a mistake; but when the image is transferred to paper, the inclination at which the mirror is held is lost sight of, and as the two sides of the larynx are symmetrical, that which really represents the right vocal cord appears as the left in the picture. The same remark of course applies to the left vocal cord, and indeed is of general application as regards the other parts on each side of the larynx. To make my meaning perfectly clear, let the reader take any laryngoscopic drawing in this book—say figure 23. Let him hold the book at the same angle that the mirror had when the drawing was taken; he will then have no difficulty in judging of right and left; and a small wart will be seen on the left vocal cord. Let him now place the book on a table (in the position in which it is usually looked at); in doing this, the plane of the paper on which the drawing is made will undergo a revolution of three-fourths of a circle. The arytænoid cartilages, which in the laryngeal mirror were farthest from, are now nearest to, the observer, and the epiglottis, which was nearest, is now farthest.

In this revolution, the image being as it were turned right over, and the two sides of the larynx being perfectly symmetrical, the right vocal cord in nature becomes the left in the picture, and the small wart on the left vocal cord appears on the right in the picture.

SECTION II.—*Practice of the Art.*

The patient should sit upright, facing the observer, with his head inclined very slightly backward. The observer's eyes should be about one foot distant from the patient's mouth, and a lamp burning with a strong clear light should be placed on a table at the side of the patient, the flame of the lamp being on a level with the patient's eyes. The observer should now put on the spectacle-frame with the reflector attached, and directing the patient to open his mouth widely, should endeavor to throw a disk of light on to the fauces, so that the center of the disk corresponds with the base of the uvula. If the observer has much trouble in projecting the light on to the fauces, he will find it convenient to incline the reflector at a suitable angle before putting on the spectacle-frame. This may be done as follows:

Taking the spectacle-frame, with the mirror attached, in the hand, and fixing the joint so that the back of the mirror is parallel with the spectacle-frame, the outer edge of the reflector should be pushed rather more than a quarter of an inch forward or back-ward, according as the lamp is on the right or left side of the patient. If the observer has chosen his position and placed the lamp as directed, on putting on the spectacle-frame, a beautiful luminous disk will appear at the back of the throat.

The patient should be directed to put out his tongue, and the observer should hold the protruded organ gently but firmly between the finger and thumb of his left hand, the thumb being above and the finger below. To prevent the tongue slipping, the observer's hand should be previously enveloped in a small soft cloth or towel, and he should be careful to keep his finger rather above the level of the teeth, in order to prevent the frænum being torn. In cases that are likely to require local treatment, the patient should be taught to hold out his own tongue, in order that the operator may be able to introduce the mirror with his left hand, while with the right he applies the remedy to the affected part.

When the observer has practiced the two first stages, he should take a small laryngeal mirror about half an inch in diameter, and after warming its reflecting surface for a few seconds over the chimney of the lamp* (to prevent the moisture of the expired air being condensed on it), should introduce it to the back of the throat. Before thus introducing the mirror, in order to prevent its being unpleasantly hot, the practitioner should test its temperature, by placing it on the back of his hand. To pass the mirror to the back of the throat with as little annoyance as possible to the patient, the following method should be adopted : The handle of the mirror should be held like a pen in the right hand, and quickly introduced to the back of the throat, its face being directed downward, and kept

* A very ingenious plan of keeping the mirror at a suitable and uniform temperature, by the aid of the electric current, has been suggested, and indeed carried out by Dr. Henry Wright. At the back of the mirror is a small shallow cell, which contains a carefully insulated loop of platinum wire; this loop is in communication with a battery of two or three cells, by means of two fine copper wires, which pass through the hollow shank and handle of the mirror. This contrivance is calculated to prove useful in the consulting-room of those much engaged in laryngoscopy; but the fact that the mirror does not become dimmed, must not be regarded as a reason for keeping it a longer time than usual in the patient's mouth. Such a procedure could only end in failure.

as far as possible from the tongue. The posterior surface of the mirror should rest on the uvula, which should be pushed rather upward and backward,

FIG. 7.

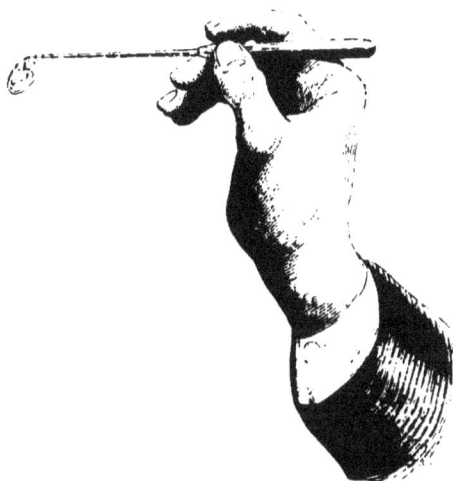

Fig. 7.—The position of the hand and mirror, when the latter has been properly introduced for obtaining a view of the larynx.

toward the posterior nares. When the mirror has been thus introduced without irritating the fauces, the observer should raise his hand slightly and direct it outward toward the corner of the mouth. This rotatory movement, which alters the inclination of the mirror, and turns its face more toward the perpendicular (while the hand is thereby kept entirely out of the line of vision), should be effected rather slowly, so that it can be arrested directly the larynx comes into view. After introducing the mirror, the observer can, if he chooses, steady it, by resting the third and fourth fingers against the patient's cheek. The exact angle which the mirror should bear to the laryngeal aperture must depend on a number of circumstances, such as the degree of flexion backward of the patient's head; the particular angle which the plane of the laryngeal aperture bears to the horizon in the case undergoing inspection; and on the direction which the ray must take to reach the observer's eye—that is to say, on the position of the observer. The practitioner should learn to introduce the mirror with either hand, for by so doing any false ideas concerning a supposed asymmetrical condition will be at once corrected; but while for the purpose of diagnosis it is very desirable to be able to use either hand, in the application of remedies to the larynx, ambidexterity is absolutely essential.

Beginners in their anxiety to get a good view often give rise to faucial irritation, by keeping the mirror too long in the patient's mouth; the same condition is also often produced by moving the mirror too much about at the back of the throat after its introduction. The practitioner should recollect that when an act of retching has once taken place, it is afterward often impossible to get a good view of the larynx at the same sitting. Moreover, the act of retching always causes considerable temporary congestion of the laryngeal mucous membrane, and thus is apt to lead the inexperienced to very erroneous conclusions. It is therefore better to introduce the mirror any number of times, keeping it in the throat only for a few seconds each time, than to let it remain longer, and thus limit the examination to one inspection. The novice must be careful to avoid touching the tongue with the mirror, for this procedure irritates the throat and spoils the reflecting surface of the mirror for the time. This can generally be avoided by keeping the back of the mirror in close proximity to, but not letting it touch, the palate. In some people, however, the uvula is in actual contact with the back of the tongue, and as in inspiration or vocalization the uvula is raised, such patients should be directed to draw in their breath, or to produce some vocal sound (such as "ah," "eh," "oh," &c.); the mirror can then be easily slipped in between the uvula and the tongue.

The difficulties solely dependent on the practitioner's want of dexterity have been already considered, but a few words must be devoted to those in part due to the patient. The obstacle may be either undue irritability of the fauces, a peculiar action of the tongue, or a pendent condition of the epiglottis. As regards faucial irritability, it is to be observed that though this condition sometimes exists of itself, it is far more often caused by the clumsiness or inexperience of the practitioner. Most patients can be examined with facility at the first sitting, and only a small proportion require any training. With timid patients—especially women—on first using the laryngoscope, it is well to place the mirror for a second on the back part of the palate, without being too particular about seeing anything. By introducing the mirror once or twice in this way, the patient's confidence is secured, and a more fruitful examination may afterward be made.

For reducing the unusually irritable condition of the fauces, the internal administration of the bromides of potassium and ammonium has been recommended; but my experience has proved the total inutility of their employment. Some advise that the patient should be directed to inhale a few whiffs of chloroform; but in those rare cases, which present much difficulty, I have found the best effects result from sucking ice for about ten minutes before the mirror is to be introduced. The most irritable fauces cannot resist this plan. The conformation of parts sometimes causes some difficulty. Thus when the tongue is drawn out, it sometimes forms an arched prominence behind, which causes trouble in introducing the mirror, and difficulty in seeing it when *in situ*. It is due to reflex action, and will be best avoided by pulling the tongue less out than usual, keeping it level with the mouth (that is to say, not holding it down toward the chin), and by cautioning the patient not to strain. Enlarged tonsils sometimes embarrass the operator. In this condition a small oval mirror should be used.

An unusually large or pendent epiglottis causes a

more serious impediment to laryngoscopy. When the valve is very large, it sometimes shuts out the view of the larynx; but the same result is more often caused by unusual length or relaxation of the glosso-epiglottidean ligaments. In the production of high (falsetto) notes, the epiglottis is generally raised, and this also happens when a person laughs; the observer will therefore do well to take advantage of these physiological facts. In a certain number of cases, however, the epiglottis remains obstinately pendent. For elevating the valve in these cases, various instruments have been invented (by Volto-lini, Bruns, Fournié, Lewin, and others), and I have myself had one contrived which has proved useful in some cases. (*See* page 48.) Most of the instru-ments hitherto invented, however, cause so much irritation, that they cannot often be employed with advantage. When the epiglottis covers the larynx in the manner described, the laryngeal mirror should be introduced lower in the fauces, and more perpen-dicularly than is usually suitable. In almost all cases, the arytænoid cartilages surmounted by the capitula Santorini can be seen, and from them we

can judge with tolerable certainty as to the mobility
of the vocal cords; the state of the mucous membrane of the larynx in other parts cannot, however,
be safely inferred from the condition of that which
covers the arytænoid cartilages.

CHAPTER IV.

THE HEALTHY LARYNX (AS SEEN WITH THE LARYN-
GOSCOPE).

IT is not intended to enter into the anatomy of
the different parts of the larynx as seen on dis-
section of the dead subject, for this is treated of in
various works on general anatomy. In other words,
the description will be confined to the internal surface of the larynx, and no mention will be made of
parts, the contour of which cannot be seen in the
mirror. The rationale of the formation of the image
has already been explained (page 59), the special
description of its individual parts will therefore be
now undertaken. In some cases, on introducing the
laryngeal mirror, only the epiglottis may be visible,
with perhaps just the tips of the capitula Santorini
at the posterior part; while in others the entire
length of the vocal cords, the ventricular bands
(false vocal cords), the small cartilages above the

glottis, the large cricoid cartilage, the rings of the trachea, and perhaps even the bifurcation of the

FIG. 8. FIG. 9.

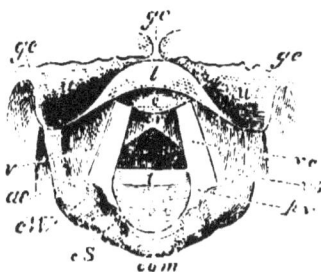

Fig. 8.—Laryngoscopic drawing, showing the vocal cords drawn widely apart, and the position of the various parts above and below the glottis, during quiet inspiration.

ge. Glosso-epiglottidean folds.
u. Upper surface of epiglottis.
l. Lip of epiglottis.
c. Cushion of epiglottis.
v. Ventricle of larynx.
ae. Ary-epiglottidean fold.
cW. Cartilage of Wrisberg.
cS. Capitulum Santorini.
com. Arytænoid commissure.
vc. Vocal cord.
vb. Ventricular band.
pv. Processus vocalis.
cr. Cricoid cartilage.
t. Rings of trachea.

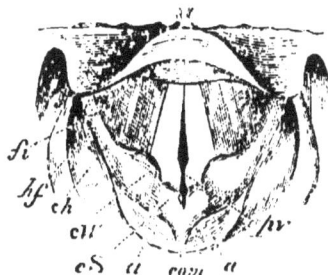

Fig. 9.—Laryngoscopic drawing, showing the approximation of the vocal cords, and the position of the various parts in the act of vocalization.

fi. Fossa innominata.
hf. Hyoid fossa.
ch. Cornu of hyoid bone.
cW. Cartilage of Wrisberg.
cS. Capitulum Santorini.
a. Arytænoid cartilages.
com. Arytænoid commissure.
a. Arytænoid cartilages.
pv. Processus vocalis.

bronchi below it, can be seen with perfect distinctness. The view varies in different cases between these two extremes.

The epiglottis varies very much in appearance in different individuals; for it may be large or small, broad or narrow, long or short. In most cases there is seen—1st, A portion of its upper surface on either side (*u*); 2dly, its free edge and a small portion of its under surface turned up in the center, and forming a kind of lip (*l*); and 3dly, another portion of its under surface, below and behind the lip, projecting as a rounded prominence,—the cushion (*c*). The upper surface is of a dirty pinkish hue; the lip is of a decided yellow color, though it has a slight shade of pink; and the cushion is invariably bright red. In some cases the whole of the under surface of the epiglottis is seen, and then it is of a bright red color. This normal coloration of the under surface of the epiglottis is apt to be mistaken (by those unaccustomed to the use of the laryngoscope) for congestion of the mucous membrane.

In some cases the epiglottis is broad (see Fig. 21, page 129), while in others it is extremely narrow (Fig. 26, page 138); and while in some people only the upper surface can be seen (Fig. 24, page 134), in others, where the epiglottis is drawn tightly to the tongue, only the under surface is visible (Fig. 21, page 129). In the center of the free edge is a slight notch (Fig. 21), which gives to the epiglottis, when seen in its entirety, its foliate appearance. But the free edge of the valve is more often turned upon itself, so that in the reflection, the notch is lost sight of, and the border appears round (Figs. 8, 9, and others).

In some cases, on account of the inclination of the epiglottis, only the profile of its free edge is visible in the mirror (see Fig. 25, page 136). In these cases the valve is represented by a thin line. Above the epiglottis, the glosso-epiglottidean folds (*g e*) may be seen, passing upward and backward to the tongue; the profile of the latter—that is to say, of its posterior superior border—is seen as a horizontal line, which, on account of the projecting papillæ, is somewhat uneven.

The ary-epiglottidean* folds (*a e*), which form the lateral boundaries of the upper laryngeal aperture, can be seen in the mirror extending obliquely downward and backward from the epiglottis to the arytæ-

---

* On account of the extreme length of the term arytæno-epiglotti-dean folds, following the Germans, I have dropped the third and fourth syllables of this unnecessarily and inconveniently long word.

noid cartilages. Near the latter are the slight pink-ish prominences of the cartilages of Wrisberg (*cW*), and a little beyond the cartilages of Wrisberg, in the same fold of mucous membrane, are two other small prominences, the capitula Santorini (*cS*), surmount-ing the arytænoid cartilages.

The cartilages of Wrisberg generally appear round, but sometimes, especially in thin people, they have a triangular shape,—the apex of the triangle being directed outward. The capitula Santorini have a roundish shape in the healthy larynx. Both these cartilages are most distinct when the vocal cords are approximated, but the clearness with which these small laryngeal cartilages can be seen, depends upon their degree of development, and also upon the amount of submucous areolar tissue surrounding them; sometimes the cartilage of Wrisberg is not to be seen at all, while occasionally there are small cartilages between it and the capitula Santorini. The breadth of the ary-epiglottidean folds varies in different people, and in different states of the larynx, appearing broad when they are relaxed, that is in inspiration, and narrow when they are tense, as in the approximation of the cords,—especially in the production of high notes. The ary-epiglottidean folds have been well described by Stœrck, as having almost the same color as the gums. The cartilages of Wrisberg and Santorini are of a rather lighter color than the rest of the mucous membrane.

The arytænoid cartilages (*a*) are usually recognized by the small cartilages of Santorini which surmount them. They can be best seen when the vocal cords are approximated. The mucous membrane covering them is generally of a rather redder tinge than that forming the ary-epiglottidean folds. Between the arytænoid cartilages is a fold of mucous membrane, the inter-arytænoid fold or commissure, which is most apparent when the glottis is widely open (Fig. 8, *com*); when the arytænoid cartilages are approximated, the commissure folds together, and is directed backward (Fig. 9, *com*). It is of a yellowish-pink color.

The ventricular bands (*vb*), commonly called the false vocal cords,* are the folds of mucous membrane

---

* *Note on Nomenclature.* The retention of the term "false vocal cords" not only perpetuates the memory of a physiological error, but makes it necessary to qualify the real vocal cords by the term "true." While a subject attracts but little attention, its nomenclature is not

which are seen below the ary-epiglottidean folds, passing obliquely in the antero-posterior diameter of the larynx, from the arytænoid cartilages to the epiglottis. They are thick, rather prominent, and of a deeper red color than the ary-epiglottidean folds. Being rather thinner, and more prominent at their lower edge (which borders on the ventricle) than else-where, this part has a lighter tint when illuminated than the rest of the ligament. When the vocal cords are approximated, a small depression may be seen near the epiglottis between the ventricular bands below and the ary-epiglottidean folds above, which I propose to call the fossa innominata (fi).

The openings of the ventricles (v) can sometimes be seen, as dark lines, between the ventricular bands and vocal cords. They are best seen in the healthy

a matter of much importance, but when, on the other hand, it excites general interest, incorrect or inconvenient terms should be carefully avoided. The upper ligaments of the larynx have, in their normal state, no direct influence on vocalization, but merely bind the ventricular orifice. In calling these ligaments "ventricular bands," I may observe that the corresponding term, "taschenbänder," is beginning to be used in Germany. In support of my nomenclature I may further observe, that if the ventricular bands were called the superior ligaments of the larynx, the vocal cords would become, by contradis-tinction, the inferior laryngeal ligaments. The mistake of giving such a name to parts having so important a function would be mani-fest to every one.

larynx of a thin subject—especially when there is
a slight disposition to spasm.

The vocal cords (*vc*), when visible, cannot be mistaken. They are seen as two pearly-white cords,
passing from the base of the arytænoid cartilages to
the angle of the thyroid cartilage. On inspiration,
they are seen almost to touch each other at their
anterior insertion, but to be separated from a quarter
to half an inch posteriorly. On phonation they
become parallel, and appear to approximate. Each
vocal cord is seen to terminate in the angle at the
base of the arytænoid cartilages, called the vocal
process (*vp*). On inspiration, this angle is directed
outward, and the glottis has a lozenge shape; but
when the vocal cords approach one another, the
angle is turned inward. This process divides the
inter-cartilaginous and inter-ligamentous portions of
the glottis.

Below the vocal cords, appears the broad yellow
cricoid cartilage (*cr*), and below it again the rings
of the trachea (*t*), are seen elevating the mucous
membrane, which between them is of a pale pink
color. Occasionally, two indistinct dark rings (the

openings of the bronchi), on either side of a bright projecting line (the angle of division between the bronchi), indicate the bifurcation of the trachea. In some rare cases a ray of light may be thrown down the right bronchus.

Though external to the larynx, it is necessary to mention the hyoid fossa (*hf*), in which foreign bodies are extremely likely to become lodged. It is bounded on the inner side by the ary-epiglottidean folds, and on the outer side by the inner surface of the thyroid cartilage. Projecting from the outer wall and sometimes forming the floor of the fossa, the greater cornu of the hyoid bone is sometimes seen glistening beneath the mucous membrane.

CHAPTER V.

ACCESSORIES OF LARYNGOSCOPY.

SECTION I.—*Auto-laryngoscopy.*

THOSE who desire to acquire dexterity in introducing the mirror at their own expense, rather
than that of their patients, or those who wish to
demonstrate their larynx to others, should learn to
employ the laryngoscope on themselves.

When auto-laryngoscopy is practiced, it is requisite that, besides the circular reflector and laryngeal
mirror, another mirror should be used : this must be
placed in such a position that the image reflected in
it from the throat-mirror can be seen by the auto-
scopist.

For practicing auto-laryngoscopy, Professor Czermak contrived a special apparatus. It has a large
reflector and quadrilateral mirror, each supported on
perpendicular bars. These mirrors are fixed about a
foot apart, and both can be turned in almost any

direction, and fixed at any height. In using this apparatus, the observer should sit at a table with the quadrilateral mirror a few inches in front of his mouth, and the reflector again a foot further back. The flame of the lamp should be near the quadrilateral mirror, the upper edge of which should be level with the lower edge of the reflector behind it. The observer now throws the light into his fauces with the reflector, introduces the warmed laryngeal mirror, and sees the image in the quadrilateral one. People facing the demonstrator can see the image in the laryngeal mirror, and those behind him in the one which he looks at. For those who wish to make accurate physiological observations, this is the best method of practicing auto-laryngoscopy.

Those who object to purchase a special apparatus can use the ordinary reflector for auto-laryngoscopy. In this case, all that is requisite is a perpendicular telescope-bar, capable of being made about a foot and a half in length, and having a broad firm base: at the top of the bar is a small projecting ball, which fits into the socket at the back of the ordinary reflector. The reflector is placed on a table, at about

eighteen inches from the observer, between whom
and the reflector there must be a small toilet mirror
or hand glass, the frame of which ought not to be
very thick. In other respects, the examination will
be conducted as already described.

A very ingenious, useful, and simple method of
practicing auto-laryngoscopy has been recommended
by Dr. George Johnson. The observer puts on his
ordinary reflector, as though he were going to
examine a patient, and sits facing a toilet mirror.
A lamp is placed on the left side of the observer,
in a line with the mirror, or slightly behind it. The
observer now, by manipulating the reflector, throws
the light on to the image of his fauces, as seen in the
toilet glass. He then introduces the laryngeal mirror into his throat, and the image of the larynx
formed on it is seen in the toilet glass, both by
the demonstrator and by persons standing behind
him. In practicing auto-laryngoscopy in this manner, the practitioner has to manage the light in the
same way as in examining patients, and he thus
learns to overcome one of the difficulties of laryngoscopy. The only disadvantage of this method, as

compared with that of Czermak, is that by it the rays of light undergo an additional reflection before they reach the larynx, and thus the image is not quite so distinct.

SECTION II.—*Recipro-laryngoscopy** (*or the Demonstration of a Patient's Larynx to others*).

In seeing patients in consultation, the laryngoscopist may often desire to have the benefit of the opinion of a colleague, who may not himself be able to use the laryngoscope; or a teacher may desire to demonstrate a patient's larynx to students. The inconveniences of showing the larynx to a third person in the ordinary way have been pointed out by Dr. Smyly, of Dublin, and he has contrived a means of overcoming them. To use his own words: "In the ordinary method, when the examiner has a full view of the vocal cords of the examinee, he calls

---

* Laryngoscopy is the term used when one person examines another's larynx. Auto-laryngoscopy means the examination of one's own larynx. Recipro-laryngoscopy expresses that particular mode of practicing the art, in which a person's larynx is reciprocal to two or more people. If any one can suggest a less pedantic, but equally correct term, I shall be ready to adopt it.

upon his colleague to view the parts; who, when he places his head beside that of the examiner, only gets a partial view—a portion of the epiglottis, one arytænoid, and perhaps a vocal cord. In endeavor-

Fig. 10.

Fig. 10.—Dr. Smyly's Recipro-Laryngoscope.

ing to see more, he pushes the examiner's head, so as to displace the light, or shakes his hand, so as to bring on nausea. Many other inconveniences will occur to the mind of the practical laryngoscopist which I shall not here allude to.

"My addition consists of a simple square piece of very good plate glass mirror, set in brass, like

the ordinary concave mirror. A second split tube
is soldered on close to the tube which exists on all
Weiss's frontal bands, and a brass rod, the ends of
which are bent in opposite directions, at an angle
of 45°.

"The mode of using this glass is as follows: The
laryngoscope is fixed, as usual, before either the
left or right eye. The brass rod is fixed in the
tube, beside that which holds the rod supporting
the reflector; and my square glass is fixed on the
other end, as is very well shown in the engraving.

"As the angles of incidence and reflection are
equal, the mirror may be turned to such an angle,
that the second examiner may be placed at such a
distance from both the patient and operator, that
his presence cannot disturb the steadiness of either.
The view the second examiner has of the larynx in
the square mirror is not inverted, being twice reflected. The right vocal cord of the examinee is
to the right-hand side of the examiner number two.

"The glass employed in the manufacture must be
as perfect and parallel as possible, so that the loss
of light may be a minimum.

"In conclusion, I may add that the additional weight of the square glass, when made in the artistic manner in which mine has been made, by Messrs. Spencer & Son, of Aungier Street, Dublin, is scarcely perceptible."—*Dublin Quarterly Journal*, vol. xxxvi, Aug. 1863.

SECTION III.—*Infra-glottic Laryngoscopy, or Tracheoscopy.**

Where tracheotomy has been performed and a fenestrated canula is worn, a very minute mirror may be introduced through the tube with its face directed upward. In this way the observer obtains a view of the larynx from below.

This method was first suggested by Dr. Neudörfer, in 1858, and was first carried out by Dr. Czermak in the following year. Since then various observers have examined patients in this way, and I have myself had an opportunity of examining several

* As it is not the trachea which is examined, the term tracheoscopy is obviously incorrect, and the expression infra-glottic laryngoscopy would more correctly describe this method of investigation.

cases. Some very interesting observations made by
a medical man on himself in this way, have been
recorded by Dr. Semeleder. This mode of examining the larynx, though of very limited application,
is extremely valuable, because it frequently happens
that, in cases where a canula is worn, the epiglottis
is bound down over the larynx by old cicatrixes,
and consequently ordinary laryngoscopy is useless.
It is well to remark that the vocal cords, when
observed from below, have a reddish color, and do
not present the peculiar white appearance which is
seen when the laryngeal mirror is placed on the
uvula.

Section IV.—*Magnifying Instruments.*

Various instruments have been invented for increasing the size of the laryngeal image, but they
are of no use in the treatment of disease. As early
as 1859, Dr. Wertheim, of Vienna, recommended
concave laryngeal mirrors for this purpose, and later,
Dr. Türck, calling attention to the fact, that the
laryngeal image is made up of a number of parts at

different distances, suggested the use of a small tele-
scope (!!) which he had fitted to his illuminating
apparatus; finally Voltolini by removing the ocular,
adapted an opera glass (!!!) which, however, he was
only able to use with sunlight.

SECTION V.—*Micrometers.*

For measuring the exact size of different parts of
the larynx, and for estimating distances, Merkel, of
Leipsig, and Mandl, of Paris, have suggested the
plan of having a scale scratched on the laryngeal
mirror. Dr. Semeleder objects to this mode of measuring, as it takes so much away from the reflecting
surface of the mirror, and he recommends that the
scale should be drawn on the frame of the mirror.
Though these scales might perhaps be advantageously employed for physiological investigations,
they are of no use to the medical practitioner.

7

SECTION VI.—*The Epiglottic Pincette.*

In a certain proportion of cases, it is impossible to obtain a satisfactory view of the larynx, on account of the pendent condition of the epiglottis. This peculiarity, which depends on the length of the glosso-epiglottidean ligaments, is probably more often congenital, but sometimes it may—to a certain extent—be due to a relaxed condition of the system generally. In cases where it does not occlude the whole of the larynx, it often hides the anterior third of the vocal cords. After inventing various instruments which did not answer the purpose, I hit upon the pincette of which the annexed wood-cut is a representation. In constructing an instrument to hold and draw forward the epiglottis, it must not only effect its end, but must do so without irritating the patient's throat. I have used the pincette in a few cases with advantage, but I would specially call the attention of inventive laryngoscopists to this subject, as I believe that a thoroughly effective instrument for drawing forward the epiglottis would be, without exception, the most useful addition to

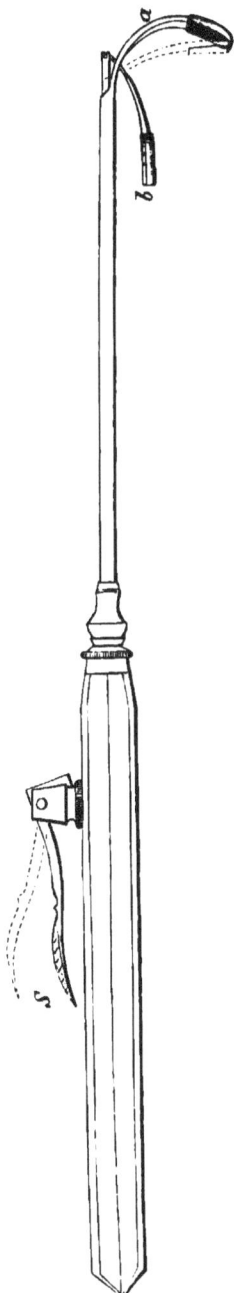

Fig. 11.—THE EPIGLOTTIC PINCETTE.

While the spring, S, is kept down, the two blades, a and b, remain widely open, and the blade, b, should come right up to the tube, which is about a quarter of an inch above it in the wood-cut. The operator passes the blade, a, behind and below the epiglottis, and draws the valve slightly forward and upward ; he then raises his finger from the spring, S, when the blade, b, advances to a, and the epiglottis is held gently between them. The blades, a and b, are a quarter of an inch broad, slightly convex from side to side (the convexity being directed forward), and covered with india-rubber.

the art of laryngoscopy. In my pincette the anterior blade curves round, so that it can be passed behind and under the epiglottis, while the posterior blade, which alone moves, opens to the extent of almost half an inch, and lies flat on the tube which contains the wire by which it is moved. The instrument is kept open by the pressure of the index finger on a spring in the handle, and when the anterior blade has been passed under the epiglottis, and has drawn it slightly upward and forward, the operator removes his finger from the spring, and the blades close and hold the valve in the desired position. The blades are broad, flat, and covered with india-rubber.

Section VII.—*The Self-holder or Fixateur.*

In applying remedies to, or operating on, the larynx, in those cases where the pincette is used for drawing forward the epiglottis, one hand is employed in introducing the instrument (brush, lancet, or forceps, as the case may be), and the other holds the pincette, so that it becomes necessary to employ some apparatus for holding the laryngeal mirror.

Fig. 12.

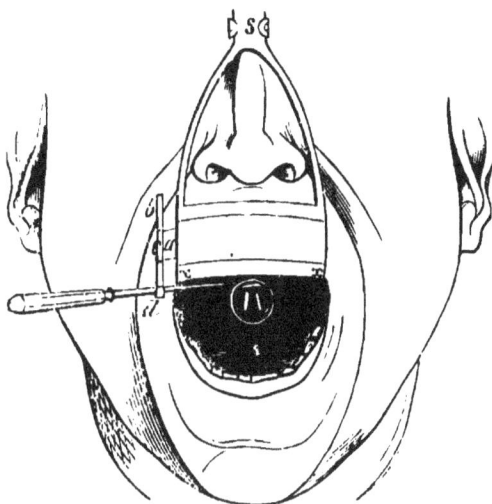

Fig. 12.—The Self-holder, or *fixateur* for holding the laryngeal mirror after introduction. A broad plate of metal rests against the upper lip, and from its lower border, *x*, a small metal plate passes backward under the upper teeth.

s. Steel spring, which passes upward and backward over the head, to below the occipital protuberance, where there is a pad.

a. Small metal plate, which can be inserted into either the right or left side of the large hollow plate, which contains it. The small plate can be drawn out to any extent desired, so that its free extremity can always be brought to the corner of the mouth. It terminates in

c, a ring which holds—*b*, a perpendicular bar that can be fixed at any height in the ring.

d. Termination of the bar in a kind of spring forceps; the blades of the forceps are very broad, and each is curved outward, so that they form a kind of groove into which the stem of the mirror easily passes, and can be as easily withdrawn. As the perpendicular bar can be made to turn round in the ring, *c*, the blades can be made to open in any direction.

Various instruments have been invented for this purpose, and one of my own design was described in the *Medical Times*, Aug. 8, 1863. I have since greatly improved this instrument, so that it can be used in any case. It consists of a mouth-piece, and a steel band which passes from its upper part, round the head, to beneath the occipital protuberance, where there is a broad pad to keep it in position. The mouth-piece is composed of perpendicular and horizontal portions; of these the perpendicular portion is two inches long and one inch broad, and it rests against the upper lip; the horizontal portion is kept in the mouth, so that the inner angle of union of the two parts corresponds to the points of the front teeth; the upper surface of the horizontal portion is covered with a thin layer of wood, to save the teeth from coming in contact with the metal. In the perpendicular portion, there is a groove passing through its entire length, and in it a steel plate runs, which has a broad ring at its extremity. Through this ring there passes a perpendicular bar, having at its extremity two broad steel blades. These blades, which open at a considerable angle, close by their own elasticity, and hold the mirror firmly. The perpendicular bar moves in the ring, so that it can be fixed at any height, and so that the blades can be turned in any direction. As the steel plate also can be drawn out of its groove, the laryngeal mirror can, in all cases, be brought as near to the corner of the mouth as desired. The steel plate can also be passed into the groove from either side, so that the laryngeal mirror can be fixed at either the right or left side of the mouth. The annexed wood-cut explains the action of the instrument, and shows the mirror *in situ*.

SECTION VIII.—*The Head-rest*.

Most people, when they are going to have the throat examined, lean back in the chair, throw up the head and open the mouth. This attitude, however, is very ill-suited for laryngoscopy, where both the head and body should be kept erect. In many cases also,—especially where the patient is at all nervous,—in applying remedies to, or operating on, the larynx, it is very desirable to be able to steady the head. For this purpose I use a head-rest, very much like that employed by photographers, except that instead of having a stand of its own, it is fixed to a chair. A strong metal plate, terminating in a

Fig. 13.

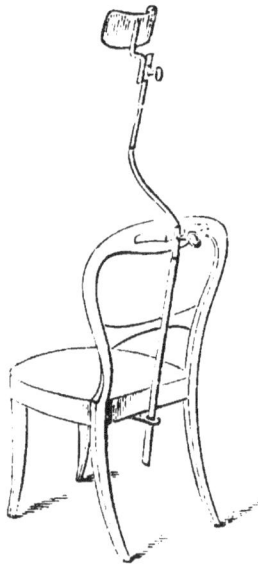

Fig. 13.—The Head-rest.

ring which projects behind the seat, is screwed to the back of an ordinary chair, to the under surface of the frame which supports the seat; and another similar projecting ring is screwed to the top bar of the chair. A strong iron bar passes perpendicularly through these rings; just above the upper ring it bends obliquely forward for about half a foot, and then again passes perpendicularly upward for another foot. This bend in the bar prevents the patient leaning back. Sliding on the perpendicular bar, is a broad, curved, semicircular pad, which supports the head, and can be fixed at any height. It allows the patient to raise his head, but prevents any movement backward or laterally. The apparatus is not unsightly, if the metal part is made of brass; and when the support is not required, the perpendicular bar and head-rest can be altogether put away. The use of the head-rest not only saves the practitioner's time; but the patient's efforts to restrain himself are greatly spared, and he consequently suffers much less from exhaustion.

CHAPTER VI.

THE APPLICATION OF REMEDIES TO THE LARYNX WITH THE AID OF THE LARYNGOSCOPE.

SECTION I.—*Solutions.*

FOR applying solutions to the larynx, squirrel's or camel's hair pencils, cut square at the end, and firmly attached to aluminium wire bent at a proper angle, will be found most suitable. The angle at which the wire is bent may vary between 90° and 120°, according as the anterior insertion of the vocal cords, or the arytænoid cartilages, have to be touched; but for most cases, an angle of 108° will be found convenient. Below the angle to the end of the brush, the instrument may measure from an inch and a half to two inches and a half, and between the handle and the angle from four to five inches. The handle should be of convenient shape and size. Great care should be taken that the brush is securely riveted to the wire. Instead of aluminium, stout copper wire coated with silver, or silver wire, may be used.

The practitioner should be provided with brushes of different sizes, and inclined at different angles. The laryngeal brush is well adapted for applying caustic, astringent, alterative, or sedative solutions to the larynx.

So many cases of acute and chronic inflammation of the larynx, successfully treated by topical remedies, have been related in the medical journals, that it is scarcely necessary to bring forward here any proofs of the value of local treatment. I shall therefore merely adduce the following two cases, to show what satisfactory results may be accomplished by very simple treatment.

Aphonia of more than two years' standing, from extensive disorganization of the vocal cords; a warty growth on the under surface of the epiglottis. Voice restored by the persevering local application of nitrate of silver.

Case 1.—Mr. W., of Sligo, consulted me in June, 1862, for loss of voice, which had existed since November, 1859. On examining the throat with the laryngoscope, the vocal cords were seen to be of a dirty gray color, and in a highly disorganized state,—their edges being serrated in a very peculiar manner; the rough toothlike processes of one vocal cord fitted into corresponding depressions in the edge of its fellow, and *vice versâ*. In the middle of the under surface, and near the edge of the epiglottis, was a small round excrescence. There was nothing of a syphilitic character in this case, and the diseased condition of the larynx seems to have originated in a severe and prolonged catarrh. In this case, I enjoyed the advantage of a consultation with Professor Czermak. He agreed with me in thinking the case a very unfavorable one for treatment,—at least as regarded the condition of the vocal cords. He also concurred with me, in recommending the local application of strong solutions of nitrate of silver. Neither of us, however, were at all sanguine as to the effect it might produce. I applied this remedy to the interior of the larynx some eight or nine times, but without any apparent effect. Mr. W. then returned to Ireland, and the same treatment was continued by Dr. Wood, of Sligo. After many months' treatment, the whisper was replaced by a very gruff voice, and when Mr. W. came over to consult me in September, 1863, the edges of the vocal cords were much more even,—the right cord being almost smooth, and the voice, though rather hoarse, was distinctly phonetic. The small warty growth had diminished slightly in size. Fig. 14 shows the condition of the larynx when the patient first came under treatment.

FIG. 14.

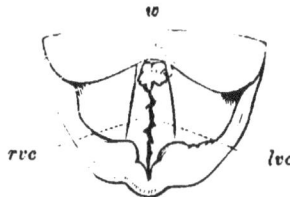

Fig. 14.—JAGGED CONDITION OF THE VOCAL CORDS, AND A WART ON THE EPIGLOTTIS.

w. Wart on the under surface of the epiglottis.
rvc. Right vocal cord.
lvc. Left vocal cord.

Warty Excrescences on the right ary-epiglottidean fold and ventricular band, destroyed by the application of solutions of nitrate of silver.

Case 2.—Mrs. A., æt. 42, from Diss, in Norfolk, applied at the Dispensary for Diseases of the Throat in April, 1863. This patient was the mother of a large healthy family. She had suffered from loss of voice for two years, but she was otherwise quite well. She attributes the aphonia to having taken cold. She had a constant inclination to clear the throat. With the laryngoscope, the aphonia was seen to depend on the presence of numerous small warty growths, situated on the right ary-epiglottidean fold and ventricular band. There was also slight congestion of the vocal cords. The appearance is shown in Fig. 15. Under the use of very strong solutions of nitrate of silver (℥ij ad ℥j), applied very frequently for some weeks, this patient became able to speak in a fairly loud, though still rather harsh, voice. When the patient was obliged to return to the country, the greater part of the excrescence had been destroyed, but a small portion still remained on the ventricular band.

FIG. 15.

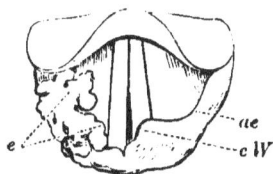

Fig. 15.—EXCRESCENCES ON THE RIGHT ARY-EPIGLOTTIDEAN FOLD.

e. The excrescences. They seem to have pushed the right cartilage of Wrisberg downward and outward and beneath the ary-epiglottidean fold, so that is not seen.

ae. The left ary-epiglottidean fold.

c W. The left cartilage of Wrisberg,

The various forms of laryngeal inflammation are for the most part analogous to similar morbid conditions occurring in other parts of the body, and the practitioner, in the selection of particular remedies, will be guided by his general experience. I shall merely remark, therefore, that among the remedies I have found most efficacious, are solutions of nitrate of silver, perchloride of iron, sulphate of copper, sulphate of zinc, carbolic acid, and iodine. Glycerine will be found a useful solvent for most of these agents. The alternation of topical remedies is often as efficacious in the cure of chronic laryngitis, as it is in the treatment of chronic inflammation of other mucous passages.

Various kinds of syringes have been invented for injecting fluids into the laryngeal cavity. I do not recommend this method of treatment, but those who wish to practice it will find my modification of Rauchfuss's injector a very manageable instrument. It is a hollow tube made of vulcanite, and suitably curved for introduction into the larynx. Near where the tube is fixed to the handle, at the upper part of the instrument, is a small hollow caoutchouc ball,

which communicates with the interior of the tube.
The injector is filled by pressing the air out of the
ball, and inserting the point of the instrument into
the solution to be used. This injector is made in
two parts, so that the same handle can be employed
with different tubes; the points of the tubes are also
made in different ways, some having a number of
small holes, so that the stream is diffused; some
with only a hole at one side, so that the fluid only
passes in one direction, &c., &c.

The principal objection to the use of injectors is,
that they have a tendency to cause more spasm
than brushes, and with them it is more difficult to
limit the amount of the application, or to restrain it
to certain spots. The injector is held between the
thumb and second finger, and the index finger
remains free to press on the ball, when the point of
the instrument has been passed into the larynx.

For the application of liquids to the larynx in the
form of a very fine spray, many kinds of *pulverisateurs* have been invented. Of these the apparatus
employed by Dr. Lewin, of Berlin, will be found
most convenient. The accompanying wood-cut ex-

plains the principle of its action. It may be advantageously used in cases of general congestion or relaxation of the mucous membrane. Its employ-

Fig. 16.

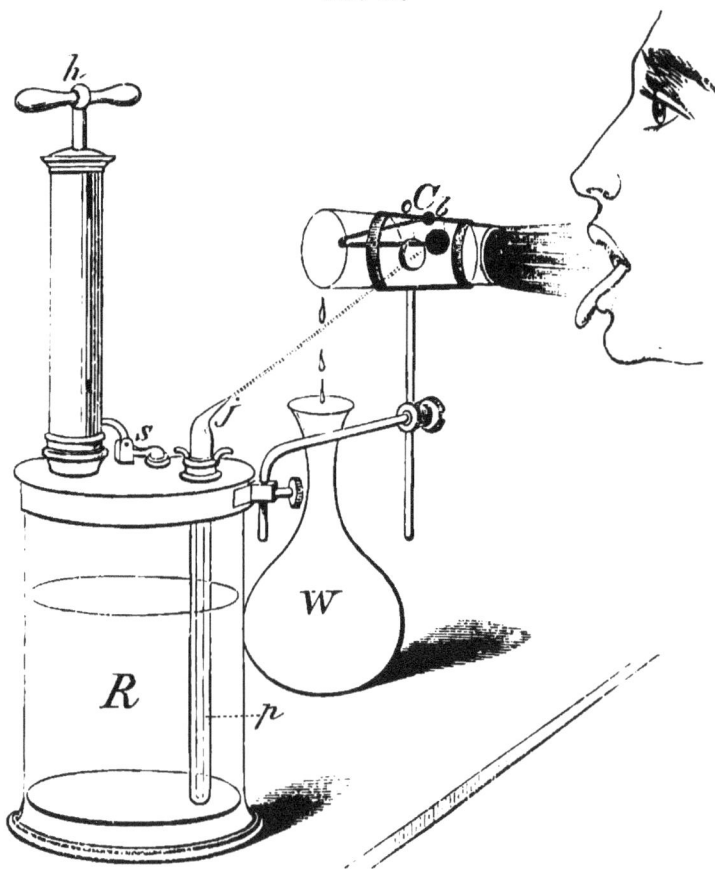

Fig. 16.—Lewin's Pulverisateur.

R, a glass receiver, into the metal top of which the air-pump is screwed. The inhaler is filled with the medicated solution by unscrewing the air-pump. Air is forced into the receiver by alternately depressing and raising the handle, *h*, with the right hand, while a finger of the left hand is kept on the extremity of the jet-thrower, *j*.

p. A fine glass pipe, which reaches almost to the bottom of the receiver, and after passing through the lid is bent at an angle of about 130°. At its extremity is a fine opening.

j. The jet-thrower, through which a very fine stream passes to the metal button *b.*

s. Safety-valve.

C. Glass cylinder, for limiting the diffusion of the spray. It slants slightly, so that the farther extremity is on rather a lower level than that near the mouth.

o. Opening in cylinder, through which the jet of liquid passes to—

b. A metal button, on which the jet breaks into a fine spray. A portion of the liquid forms drops, which run into—

W. The waste-bottle.

The patient's mouth should be placed close to the end of the cylinder, and the tongue protruded.

ment in throat affections is more particularly indicated in cases where, from circumstances, the patient cannot visit his medical attendant sufficiently often, and is thus obliged to carry out the treatment himself. Weak solutions of carbolic acid, tannin, and perchloride of iron, have been advantageously employed by me in this way. I do not recommend the use of these *pulverisateurs* for the inhalation of caustic solutions.

SECTION II.—*Powders.*

Powdered substances may be introduced into the larynx in various ways; but either the apparatus of Dr. Fournié,* or my modification of Rauchfuss's injector (already described), will be found the most convenient instruments. In some cases of laryngorrhœa, the employment of alum in this way will be found useful.

SECTION III.—*Solid Nitrate of Silver.*

For applying the solid nitrate of silver to the larynx, the only instrument which is thoroughly safe, and at the same time easy to use, is the "laryngeal cauterizer." It consists of a piece of aluminium wire, bent at the same angle, and of the same length above and below the angle, as the laryngeal brush. The wire is roughened at its extremity, and then dipped into some nitrate of silver fused over the spirit-lamp. In this way a certain quantity of the nitrate adheres firmly to the wire. An ingenious

* Dr. Fournié's instrument may be obtained of Charièrre, of Paris.

porte-caustique has been invented by Fauvel, in which, while the stick of nitrate of silver is safely inclosed, the point, by a spiral spring behind it, is always kept protruding. Dr. Stoerk, of Vienna, also, when laryngoscopy was quite in its infancy, contrived a porte-caustique, in which the caustic remains concealed till brought to the part desired to be touched, when, by pressure on a spring in the handle, it is made to protrude.

My laryngeal lancet (see page 115) is provided with a small piece of aluminium wire, which can be fitted on in place of the cutting-blade : in this way it becomes a guarded porte-caustique. The nitrate of silver is attached to the wire by fusion in the way already described.

Besides these instruments, various others have been invented; but the simple aluminium wire, which I am in the habit of employing, answers the purpose perfectly well. The solid nitrate will be found useful for touching ulcers, condylomata, and the base of growths after evulsion has been practiced.

Section IV.—*Escharotics*.

If to the stock and tube of my laryngeal lancet, a piece of aluminium wire, roughened at its extremity, is fitted on, in place of the cutting-blade, escharotics can be applied without danger, and often with great benefit. Instead of the duck-billed tube which fits on at the joint below the angle, a large silver tube should be adjusted, so that there is some space between the aluminium wire and the inner surface of the tube. For applying escharotics, I have also used a simple glass brush, firmly fixed to the end of a piece of curved aluminium wire. The brushes, however, do not answer well, as the fine glass hairs, though they do not break, are apt to come out. Where the greater part of the mucous membrane of the larynx is covered with vegetations, as not unfrequently happens, it is useless to attempt to remove them by the mouth, and foolish to open the larynx (after the manner of Ehrmann). In these cases the greatest benefit may result from the use of escharotics; and I have, on different occasions, applied nitric and chromic acids, Vienna paste, and a mixture of caustic soda and lime. The most satisfactory results have followed the use of the last preparation. This class of remedies should only be employed by those who have had much practice in introducing instruments into the larynx. The happy effect of escharotics is illustrated by Case 6, page 128.

Section V.—*Galvanism*.

By a very simple instrument of my contrivance, the electric current can be applied directly to the vocal cords. The important feature in the laryngeal galvanizer is, that the current does not pass beyond the handle, till the sponge is in contact with the vocal cords. The instrument is held in the hand, between the thumb and second finger, and when the sponge is in contact with the vocal cords, the operator with his index finger presses on the spring in the handle, and the electric current passes through the larynx to the skin externally. By placing the sponge of the galvanizer on the arytænoid cartilages, both branches of the pneumogastric nerve are stim-

Fig. 17.

Fig. 17.—THE LARYNGEAL GALVANIZER.

The instrument is connected with an electric machine by the wire *W*, which is attached to the metal ring *R*. When the ivory handle *I* is pressed upon, the metal spring *S* connects the two rings *A* and *B*, and the current passes to the point *P*, which is covered with sponge. The wire between *P* and *A* is contained in caoutchouc tubing, and forms the rod *R*. The instrument is held by the glass handle *G*.

ulated. My instrument is now extensively employed in France, Germany, and this country; and Drs. Smyly, George Johnson, Fauvel, Tobold, and others, have borne testimony to its value. Its employment is indicated in functional aphonia, and in most cases of vocal weakness, where there is no structural disease.

The source of electricity is not a matter of any importance, but its application to the vocal cords will be facilitated by the patient wearing a kind of elastic necklet, in the center of which is a piece of metal covered with sponge. This plate of metal, which is inclosed in cotton, is about three inches long, and one and a half broad, and is bent back in the center, so that, when applied, it corresponds to the thyroid cartilage. Projecting forward from the center of this thyroid pad is a metal eye, by which it may be connected with the electric machine. The pad should be wetted before it is put on the patient's neck. The employment of this necklet enables the operator to dispense with assistants. When the point of the galvanizer is placed on the vocal cords, the electric current passes right through them in all directions, to reach the pole over the thyroid cartilage.

The following two cases illustrate its use :

Complete Loss of Voice of two years' standing cured by one application of Galvanism.

Case 3.—Miss T., æt. 26 years, was sent to me March 7, 1864, by Dr. C. J. B. Williams. For the last two years her voice had been entirely suppressed, and she had only been able to speak in a feeble whisper. The aphonia originally came on with cough and cold, and when I saw her, she was rather delicate, had a bad appetite, and was easily fatigued. Her medical attendant in the country had tried local and constitutional treatment, and various tonics had also been prescribed for her by Dr. Williams, but all without effect. On examining this patient with the laryngoscope, and finding the vocal cords perfectly healthy, though relaxed, I passed the electric current through them. The voice was immediately restored. I saw the patient once or twice afterward, but there was no relapse.

Loss of Voice of three years' standing cured by two applications of Galvanism.

Case 4.—The Rev. Henry A., æt. 31, suffering from pulmonary phthisis, consulted me in September, 1863. He had suffered from weakness of voice for six years, and for the last three had not been able to speak above a whisper. He was much emaciated, and there was a small cavity at the apex of the left lung. It was supposed that he had laryngeal, as well as pulmonary, consumption; but with the laryngoscope the mucous membrane of the larynx was seen to be very pale, and covered with thin frothy mucus. On attempted phonation, the vocal cords scarcely moved at all. The first application of galvanism produced no effect; but on its repetition, a shrill feeble sound was uttered. After this time, without any further application of galvanism, the voice gradually became stronger, and at the end of October it was more powerful than it had been for many years. After the restoration of the voice, the general health became greatly improved.

For further illustrations of the value of the direct application of galvanism, I must refer to my pamphlet on the *Treatment of Hoarseness and Loss of Voice by the Application of Galvanism to the Vocal Cords.* (London: T. Richards.)

CHAPTER VII.

OPERATIONS ON THE LARYNX.

SECTION I.—*Scarification and the Opening of Abscesses.*

FOR scarifying the mucous membrane of the larynx in acute or chronic œdema of the larynx, for opening abscesses, and, in some rare cases, for dividing laryngeal growths, I have contrived an instrument which, in many instances, has proved very serviceable. It consists of a small double-edged knife or lancet, which is contained in a hollow tube, suitably curved for introduction into the larynx. The point of the lancet is concealed in the duck-billed extremity of the tube, till forced out by pressure on a spring in the handle. The stock of the instrument is provided with tubes bent at different angles, and below the angle is a joint which enables the operator to lengthen or shorten the tube. This arrangement allows for the varying inclination which the plane of the laryngeal aperture bears to

Fig. 18.

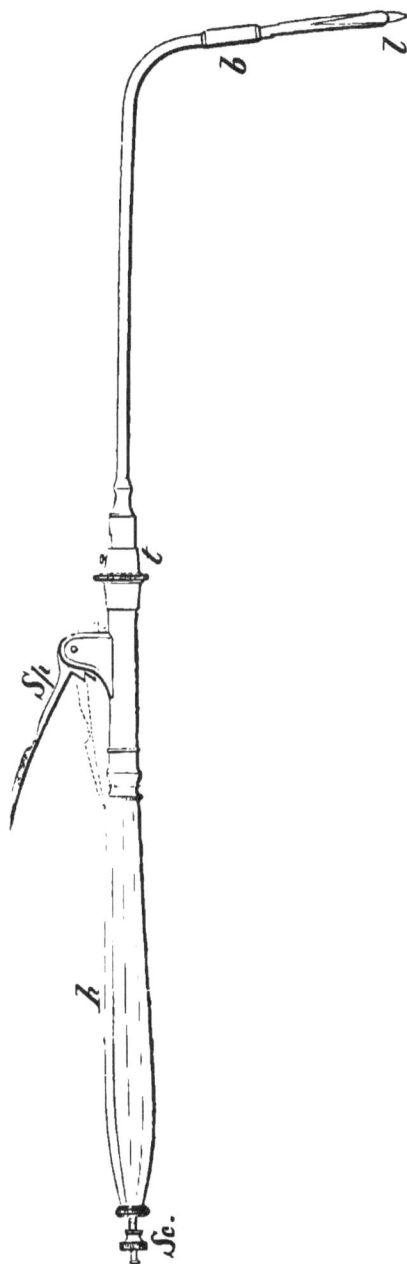

Fig. 18.—THE LARYNGEAL
LANCET.

Sp. The spring which forces
out the lancet: when it is
pressed down to the dotted
line, the lancet *l* protrudes.

h. The handle—the same as
that used for the forceps.

Sc. The screw, by turning
which, the length of the
point of the lancet can be
regulated.

t. Junction of the barrel and
stock of the instrument. At
this point, barrels curved at
different angles, can be applied. This part of the instrument is thicker than it
need be, when each instrument (forceps and lancet)
has a separate stock. To
the left side of *t*, a small
disk has to be inserted, which
fixes the tube, and allows
the chain inside to move.
In the forceps, on the other
hand, the chain is fixed, and
the tube moves.

b. The bayonet joint. A shorter or longer tube can be put
on here, according to circumstances, and the blade can
be taken out and cleaned.

the horizon, and renders the lancet fit for operating either at the upper or lower part of the larynx. The length of the blade is regulated by a screw in the handle. The instrument is held between the thumb and second finger, and when its extremity is brought opposite the part which the operator wishes to lance, he presses on the spring in the handle with his index finger.

The principal use of this instrument is in œdema of the glottis, but it may be employed for puncturing cystic tumors. A very interesting case of this sort occurred to Mr. Durham;* and Bruns, in his

* The following is the description of the case:

"On making a laryngoscopic examination, the epiglottis could not be distinguished in its normal form, but instead there appeared a large round tense tumor, projecting backward and downward, and completely covering in and concealing the glottis. On either side, and rather behind this, portions of the arytæno-epiglottidean folds could be seen swollen and apparently œdematous. The tumor could be just reached by the finger. Feeling certain that it contained fluid, Mr. Durham, with the concurrence of Dr. Wilks, at once proceeded to make an incision into it, by means of a long, curved, sharp-pointed bistoury, partially surrounded with sticking-plaster. The incision was followed by a sudden gush of thick glairy mucus, mixed with a little pus and blood, which, on subsequent examination, proved to be precisely similar to the contents of a ranula beginning to suppurate. All the patient's symptoms were at once relieved, and in the evening he was singing in bed. In the course of a few days he was perfectly well. Examinations were made from time to time, and it was interesting to watch the gradual subsidence of the œdema, and the return of the parts to their normal condition. The patient was examined second case of laryngeal growth, used a curved bistoury for dividing its base.

Chronic Œdema of the Right Ventricular Band (causing great difficulty of breathing, hoarseness, and pain), cured by scarification.

Case 5.—Charles C., æt. 22, applied at the Dispensary for Diseases of the Throat, May 4, 1863, on account of great difficulty of breathing, hoarseness, and pain in the throat. He had suffered since March, 1861, and for more than a year he had never been able to lie down at night. When he did get to sleep (in an arm-chair) he often woke with the most distressing dyspnœa, and said he felt as if he should be strangled. He had attended at the Middlesex, Brompton, and other hospitals. On making a laryngoscopic examination, the right ventricular band and aryepiglottidean fold formed together a large tumor which projected across the glottis, and concealed from view the anterior two-thirds of the left vocal cord. The swelling was of a deep purple-red color. The mucous membrane over the arytænoid cartilage was also inflamed and swollen. The case was diagnosed to be one of chronic œdema of the larynx, and was freely touched with a strong solution of nitrate of silver. This treatment was continued every

nearly four months after the operation; he was in every respect perfectly well. There was no appearance of the cyst (for such evidently was the nature of the tumor), but the cicatrix of the incision could be just distinguished on the lower part of the laryngeal aspect of the epiglottis." (*Med. Times and Gazette,* Nov. 21, 1863.)

I have particular pleasure in calling attention to this very interesting case, as, when it was brought before the Medico-Chirurgical Society (Nov. 10, 1863), some observations which I made at the time were misunderstood.

other day for a month, with little benefit to the patient; indeed, though the œdema did not increase, the patient became weaker, and the voice was completely extinguished.

FIG. 19.

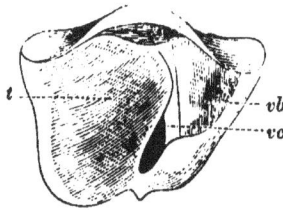

Fig. 19.—CHRONIC ŒDEMA OF THE LARYNX.

t. A large semi-transparent tumor formed by the right ary-epiglotti-dean fold and ventricular band. It projects across the glottis and eclipses part of the left vocal cord.

vb. The left ventricular band.

vc. The left vocal cord.

June 8th.—I scarified the œdematous swelling, and after the operation, the patient expectorated a considerable quantity of blood and frothy fluid.

June 10th.—On examining the larynx, the swelling did not appear much diminished. I again lanced the part freely. The next day the patient was greatly relieved; he had slept well for some hours, and woke refreshed and comfortable—a pleasure that he had not known for more than two years. The laryngoscope showed that the swelling had gone down very much, and the right vocal cord was now seen to be rather congested. I again scarified the part.

June 15th.—The patient said that he felt quite well, and asked if he could return to his work. Scarcely a trace of the œdema remained; but there was still a slight abnormal projection situated posteriorly over the right arytænoid cartilage, and the mucous membrane of the larynx generally was also rather redder than in the normal condition. The voice was at first a little hoarse, but soon became natural, and the respiration was free from any embarrassment. This patient occasionally calls at the Dispensary just to show himself, but he is perfectly well, and has been so ever since his larnyx was lanced.

SECTION II.—*The Extirpation of Growths and the Removal of Foreign Bodies from the Larynx.*

The extirpation of growths from the larynx, which was at one time[*] scarcely heard of, is now by no

[*] The only cases which I have been able to find in pre-laryngoscopic times are the following:

1st. It appears that a certain Koderik once successfully operated on a case of laryngeal growth, with a curved flexible instrument (rosenkranzartig). Nothing further is known of this case. (Semeleder, p. 50.)

2dly. Pratt performed sub-hyoid laryngotomy, for the removal of a tumor, which grew on the left half of the under surface of the epiglottis, and which, though it projected into the fauces, could not be got at from above. A firm, fibrous, grayish-white growth was extirpated. The case did very well. (Semeleder, p. 60.)

3dly. Sir Astley Cooper, with his finger, removed a large cancerous tumor, about the size of a hen's egg, from the under surface of the epiglottis. It grew again, and was again removed, and the patient finally died from haemorrhage. The specimen is preserved in the Museum of Guy's Hospital (No. 1685).

4thly. Ehrmann removed a growth from the left vocal cord in the following way: Tracheotomy was first performed by dividing the cricoid cartilage and several of the upper rings of the trachea; after the patient had had a respite of forty-eight hours, the larynx was

means an uncommon operation. Evulsion, with forceps of suitable construction, is the best mode of

divided in the median line, up to the base of the hyoid bone. When
the two halves of the thyroid cartilage were drawn apart, the growth
was seen on the left vocal cord, and removed with the knife. The
patient recovered from the operation at the end of three weeks, *but
the aphonia remained;* he unfortunately died five months later from
typhus. (*Histoire des Polypes des Larynx.* Strasbourg, 1850.)

5thly. Dr. Horace Green removed a pedunculated tumor (about the
size of a cherry) which was (thought to be) attached to the left vocal
cord. When the mouth was widely opened, and the patient coughed,
a round white fibrous-looking tumor could be seen projecting upward
between the ary-epiglottidean folds. Green succeeded in seizing the
growth with the ordinary tonsil forceps, and then in dividing it with
a long slender knife. (*Polypi of the Larynx,* p. 56. New York,
1852.)

6thly. Professor Middeldorpf, of Breslau, succeeded in removing a
tumor from the upper opening of the larynx, by means of the galvano-
caustic wire. "The sarcomatous growth showed a high degree of
cell-development," and as a portion remained behind, a very doubtful
prognosis was given: solutions of nitrate of silver were afterward
used. Rühle, who saw the case six years after the operation, states
"that there was no symptom at that time of any return of the
growth." (*Galvanokaustik,* p. 212, and Rühle, p. 229.)

From an analysis of the five latter operations (and the first one is
so vague that it must necessarily be excluded), it appears, that in
those cases where the growths were removed by instruments introduced through the mouth, they could all be seen, and in two instances
(the cases of Sir Astley Cooper and Professor Middeldorpf) could be
felt with the finger. In Dr. Green's case the tumor could be seen, and
though it was thought to be attached to the vocal cord, it more probably grew from the ventricular band or ary-epiglottidean fold. If
the polypus had been attached to the vocal cord, it could not have
been seen projecting through the opening of the larynx, unless it had
been unusually large, or its pedicle had been much longer than is
usually the case. Neither of these conditions appears to have existed.
In the cases of Ehrmann and Pratt, the operations were indirect, and
preceded by tracheotomy. Since "the eye learned to direct the hand"

removing laryngeal growths. The forceps which I use is contained in a tube, and its teeth are made to approximate by the passage of the tube over the shoulder of the blades. In seizing growths, the extremity of the instrument, therefore, scarcely moves at all. The tube which contains the forceps is made of steel, and has a diameter of one-tenth of an inch; it is bent at an angle of 110°, but to the same stock, barrels of different angles can be applied. Just below the angle is a joint, which enables the practitioner to clean the forceps, and apply shorter or longer blades, as the case may require. The spring which forces the tube over the forceps is at the anterior and upper part of the handle; and the

to the interior of the larynx, an immense number of cases of laryngeal growth have been successfully removed. Professor Bruns, of Tübingen, was the first to operate in this way; and since then Lewin, Fauvel, Semeleder, Tobold, and others on the Continent, with forceps of various construction, have extirpated a great many laryngeal tumors. In England, Dr Walker, of Peterborough, was the first who succeeded in removing a laryngeal growth. He used a modification of Gooch's double canula, which he called an *écraseur*. (*Lancet*, November, 1861.) At a later period Dr. Gibb, who still further modified this instrument, reported several cases of laryngeal growth treated in the same way. Dr. Russell, of Birmingham, has also recorded a very interesting case under his care, in which a laryngeal growth was removed by Messrs. Bracey and Bolton, with an ordinary pair of curved forceps. (*Dr. Russell on Laryngeal Disease*, p. 16. London: 1864.)

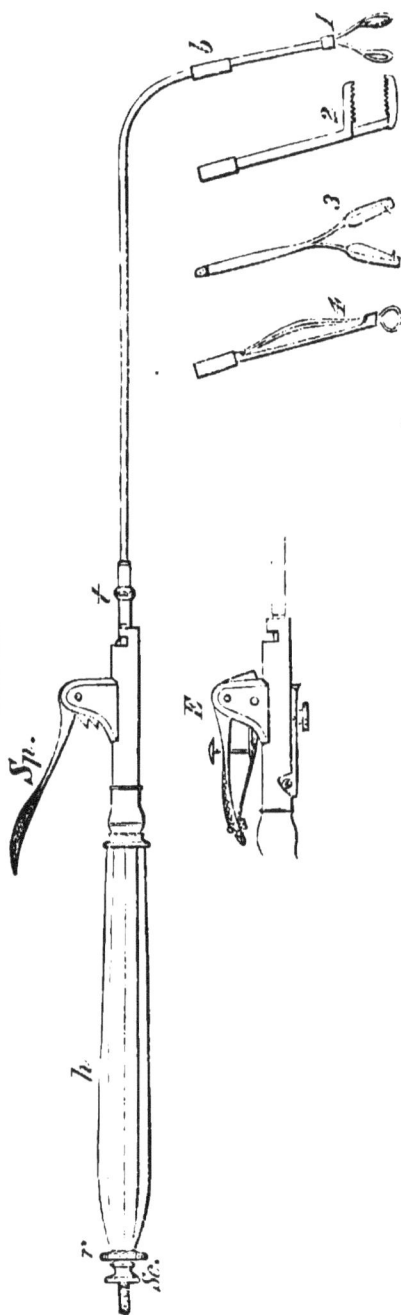

Fig. 20.

Fig. 20.—THE LARYNGEAL FORCEPS, SCISSORS, AND ÉCRASEUR.

Sp. The spring, by pressing on which, the tube is forced over the base of the forceps.

t. The junction of stock and barrel. At this point, tubes bent at different angles can be applied.

b. The joint at which longer or shorter tubes may be applied, and the blades taken out and cleaned.

h. The handle.

r. The ring, by turning which the forceps revolve, so that the blades open in any direction.

Sc. Screw for taking the instrument to pieces, cleaning it, &c.

1. The perpendicular blades.

2. The horizontal blades.

3. The scissors with hooks attached to them.

4. The *écraseur.* The loop can, if desired, be about four times the size that it appears in the drawing.

E. The modification of the stock, which is required when the *écraseur* is combined in the instrument.

* Both the forceps and lancet were originally made for me by Mr. Krohne.

operator, holding the instrument between his thumb
and second finger, presses on the spring with his
index finger.   At the posterior part of the handle is
a ring, by which the forceps can be made to revolve,
and in this way the blades can be made to open
backward and forward, or from side to side.   This
arrangement enables the operator to seize excrescences, whether they grow from near the anterior
insertion of the vocal cords, the arytænoid cartilages,
or either side of the larynx.   The blades of the
forceps have sharp cutting teeth all round their
edges.   For most cases, the blades which pass down
perpendicularly from within the tube which contains
them, are convenient; but sometimes, where the
growths are thin, membranous, and have an extensive origin from the side of the larynx, a forceps, the
blades of which open horizontally, will be found more
convenient.   In this case the forceps has in fact only
one blade, which is at right angle to its shank, the
other blade of the forceps being let into the tube:
the two blades of the forceps close when the tube
containing the upper blade is forced down, by the
pressure of the index finger on the spring in the

handle. When the growth has an extensive lateral origin, it is apt to be pushed on one side by the blades of the perpendicular forceps. Under these circumstances, the lower blade of the horizontal forceps can be passed beneath the growth, and the upper one is then forced down on the top of it.

To recapitulate : the advantages of this instrument are, first, that it can be made of any length ; secondly, that it can be inclined at any angle ; and, thirdly, that the blades can be opened in any direction. The use of this instrument is illustrated a little further on (pp. 128 to 138).

At the joint below the angle of the instrument just described, instead of the forceps, scissors can be fitted. In order that the blades should cut well and easily, the shanks of the scissors should cross one another above the blades ; the scissors which I use for cutting through growths have hooks on each blade, which seize the divided particles, and prevent their falling into the trachea. I have sometimes employed scissors also for cutting through cicatrixes : in this case the hooks are not required. (See Case 13, p. 139.)

It is to be noticed that in my forceps the tube passes over the blades, while in the lancet the tube does not move, but the blade advances. By a most ingenious arrangement* the same stock has been made to answer for both forceps and lancet.

At the joint below the angle, an *écraseur* can also be applied in place of the forceps, though this addition requires the insertion of an additional mechanism below the spring (see Fig. 18, *E*). By each pressure of the index finger on the spring in the handle, the wire is drawn through the ring, at the extremity of the instrument, and two or three movements of the finger are sufficient for the purpose. The *écraseur*, however, though it may be used for removing growths from the epiglottis,—but even for that purpose it is obviously inferior tho te forceps,— is not at all adapted for operating on tumors in the interior of the larynx. Professor Bruns, of Tübingen, in his first case, tried the principle of the wire loop, but soon perceived that it was not applicable to the larynx.†

In order that a growth could be removed from

* This was effected at my request by Mr. Mayer, of Great Portland Street, who has also combined the *écraseur* in the same instrument.

† Brit. and For. Med.-Chi. Rev., No. LXVII, p. 133. July, 1864.

the larynx with the *écraseur*, it would be necessary, not only that it should be distinctly pedunculated, but also that the peduncle should be firm and short (otherwise the tumor would hang down, and the wire could not be slipped over it). These conditions scarcely ever exist, and my own experience, as well as that of Türck, Lewin, and most Continental laryngoscopists, clearly proves that growths in the larynx occur much more frequently as warty excrescences than as true polypi. Indeed, out of thirty-one cases of laryngeal growth that I have inspected with the mirror, in only one (Case 10) was the tumor really pedunculated; even in that instance, the presence of the peduncle was inferred, not seen. In by-gone times, before the larynx was an organ of especial interest, these small vegetations were generally overlooked by pathological anatomists, and only the larger tumors attracted notice. In spite of this circumstance, however, there are not more than three or four distinctly pedunculated growths in the museums of the London hospitals.* Again, in

---

* In one or two of the cases described as "pedunculated," the foot-stalk is as broad as any other part of the tumor. Thus, in specimen No. 54 of the W series in St. Thomas's Hospital Museum, the growth which is described as pedunculated is about the size of a small walnut,

those few cases which are pedunculated, the stalk
does not remain firm when the growth attains to any
size ; but, on the other hand, the tumor falls down
from its own weight, and drags the peduncle with it.
It thus becomes impossible to get the wire round
the stalk. It would be easy to bring forward numerous other objections to the use of the wire loop,
but the probability of pieces falling into the trachea
is alone sufficient to condemn the instrument. Occasionally a particle may get entangled in the wire, or
may be jerked out of the larynx by it; but, as a rule,
portions, divided by the thin wire of the *écraseur*,
cannot do otherwise than obey the laws of gravity,
and fall into the trachea. The danger of thus allow-
ing particles to fall down the windpipe has been
pointed out by Lewin, and it can be readily appreciated by any one. If large, they would be likely to
cause fatal dyspnœa; if small, they would be almost
sure to give rise to the formation of abscesses, and

and the pedicle is attached to the entire length of the ary-epiglottidean fold. The only two pathological specimens in our museums, in
which an *écraseur* could possibly have been used, are those referred
to by Ryland, in King's College Hospital Museum. Had they been
divided by a wire loop, they would almost inevitably have caused
death, by falling into the trachea.

thus very likely to phthisis. The *écraseur*, therefore, if used at all, should only be employed in cases where the growth is situated on the lip or upper surface of the epiglottis,—in other words, only in cases where the tumor can be reached with the finger.

Five large spongy Excrescences in the Larynx; one on the under surface of the epiglottis; another on the right ventricular band; a third on the left ventricular band; a fourth on the left vocal cord; and a fifth on the right vocal cord, and the mucous membrane below the cord. The four upper excrescences were removed with the forceps.

Case 6.—William W., æt. 44, applied to me April 10, 1863, on account of loss of voice. He stated that his general health was very good, but that three years ago he had caught a cold and bad sore throat, and since then he had not been able to speak a word out loud. At Christmas his breathing was much affected, and he thought he should have been suffocated; but the attack passed off, and he said that, with the exception of not being able to speak out loud, he was now quite well. He had never had syphilis. On making a laryngoscopic examination, the laryngeal mucous membrane, above and below the vocal cords, was seen to be covered with dark reddish spongy excrescences. One was situated on the right side of the under surface of the epiglottis, another involved the whole right ventricular band, a third covered the whole of the right vocal cord, a fourth occupied half of the left ventricular band, and a fifth the anterior half of the left vocal cord.

Below the right vocal cord a number of smaller excrescences were also seen extending down into the trachea.

The appearance is shown in Fig. 21. This case was seen by Drs. Czermak, Frodsham, George Johnson, Wahltuch, and others. With my laryngeal forceps, I succeeded, in a number of sittings, in removing, in small fragments, the whole of the four upper excrescences. This included the one seated on the left vocal cord. These fragments were kindly examined for me by Dr. Andrew Clark. He "found them to consist of numerous yellowish, hard, nodular, or warty-looking particles. Under the microscope some of

Fig. 21.

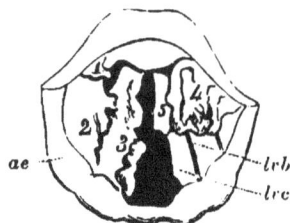

Fig. 21.—Excrescences in the Larynx.

1, 2, 3, 4, 5. Separate growths on the epiglottis, right ventricular band, right vocal cord, left ventricular band, and left vocal cord.

ae. Ary-epiglottidean fold.
lvb. Left ventricular band.
lvc. Left vocal cord.

these masses consisted entirely of enlarged racemose glands, the terminal vesicles of which were filled with minute nucleated cells and granular matter. Others were true papillary growths, consisting of more or less perfect connective tissue, clothed with many layers of epithelium, the outermost layer of which was in a state of partial desquamation. A few of the papillæ were either quite hollow, or had contained fluid." He regarded the case as one of "Granular Wart." The small particles which were torn away with the forceps produced so little effect on the bulk of the large growth on the right vocal cord, that I was induced to try the effect of escharotics. Nitric acid and chromic acid were both applied several times with decided advantage, but the greatest benefit resulted from the employment of a mixture of caustic soda and lime. The growth was reduced to a quarter its former size, and the patient has recovered a loud and tolerably clear voice. This patient is still under observation.

Warty Excrescences on and beneath both the vocal cords (causing loss of voice of four years' standing) removed with the forceps.

Case 7.—Mrs. A., aged 35, applied at the Dispensary in April, 1863, though in consequence of my absence from town she did not come under my care till the following month. I had previously (in December, 1862) seen the patient, at Mr. Maunder's request, in conjunction with Dr. Gibb, and the latter author has referred to the case at page 156 of his work, and has also given a rough sketch of the laryngoscopic appearance. The patient stated that she caught cold in 1859, was very hoarse for two years, and that in 1861 her voice had become quite suppressed. For the last two years, she had always spoken in a whisper. There was no history nor symptom of syphilis or phthisis. With the laryngoscope, both vocal cords were seen to be of a dirty grayish color, and in an irregular papillomatous condition: the appearance is shown in Fig. 22. Subsequently I discovered two growths,—one below each vocal cord. As the diseased condition of the cords was so general, and the growths on the cords were so imperfectly developed, I thought that the case would be most easily treated by caustics. Strong solutions of nitrate of silver were accordingly applied, but they produced so much dyspnœa, that the treatment was obliged to be discontinued.

Under these circumstances, I tried to use the forceps; but the patient being unable to open her mouth widely, the laryngeal aperture being exceedingly small, and the growths on the vocal cords most minute, great difficulty was expe-

FIG. 22.

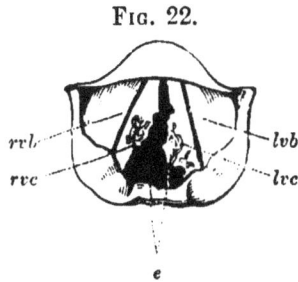

Fig. 22.—PAPILLOMATOUS EXCRESCENCES ON, AND BENEATH, THE VOCAL CORDS.

rvb. Right ventricular band.
rvc. Right vocal cord.
lvb. Left ventricular band.
lvc. Left vocal cord.
　e. Irregular papillomatous excrescences covering the vocal cords.

rienced, and it was only after repeated failures that I ultimately succeeded in clearing the vocal cords of the warty growths which covered them. The growths below the cords, which afterward became distinctly visible, being of larger size, were removed with much less difficulty. A month after the removal of the last growth, the patient's voice was completely restored. I have not seen her now for some time, but I lately received a note (dated October 31, 1864) from Mr. Brown, of Finsbury Circus, who sent the patient to me, in which he says, "I called on Mrs. A. this evening, and am pleased to find her voice is entirely restored by your treatment."

Loss of Voice of nine years' standing, caused by a small excrescence on the left vocal cord; the warty growth was removed with the forceps, and the voice completely restored at the end of a month.

Case 8.—Henry R., æt. 45, a gas-fitter, applied at the Dispensary, May 1, 1863, on account of loss of voice of nine years' standing. He stated that he had attended at various metropolitan hospitals, and had lately been at the Brompton Hospital. On examining his throat with the laryngoscope, a small round excrescence, about the size of a pea, was seen on the left vocal cord. The warty growth was situated on the free edge, and exactly in the middle of the cord, and on attempted phonation it was seen that, owing to the projection of the growth, the cords could not become approximated. On the right cord, exactly opposite

FIG. 23.

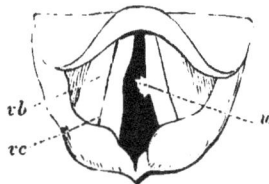

Fig. 23.—A SMALL WART ON THE RIGHT VOCAL CORD.

vb. Right ventricular band.
rc. Right vocal cord.
w. Wart on left vocal cord.

to the wart on the left cord, there was a distinct round indentation. The laryngoscopic appearance is seen in Fig. 23. I had the opportunity of exhibiting this patient to Drs. Czermak, Wahltuch, and others. There was some difficulty in removing this growth, owing to its small size, and the more than usual awkwardness of the patient, and it was not till the fourth sitting that it was successfully seized and removed. Dr. Andrew Clark examined microscopically the portions removed with the forceps, and the following was his report: "The growth was found to consist of two sets of particles, one membranous, the other warty or obscurely papilliform. The membranous portions consisted of from twenty to thirty layers of scaly epithelium, surrounded and penetrated by a confervoid growth. The epithelial cells composing the layers were polygonal, flattened, nucleated, and easily affected by weak alkalis and acids. The nucleus of each cell was oval, abruptly defined, rather large in proportion to the containing cell, in most cases surrounded by a clear halo, and in some showing signs of division. The papillary portions consisted of simple outgrowths of nucleated connective tissue and rudely-formed blood-vessels, clothed with numerous layers of scaly epithelium, similar to those already described. Some of the papillæ exhibited large vacuoles or spaces filled with colloid matter, which, in one or two instances, had burst through the covering epithelium." Dr. Clark considered the tumor to be a true wart. Immediately after the operation the patient spat up a few teaspoonfuls of blood, and the same day he was able to sound his voice. The next day he complained of a feeling of great soreness, and there was so much involuntary objection to a laryngoscopic examination, that I was unable to see exactly how the wound looked. Nine days later, however, the mucous membrane over the left vocal cord, where the growth had been, looked rather puckered, and the depression on the right cord was still visible. At the end of a month, the voice was perfect, and all morbid appearance in the larynx, including the little pit on the edge of the right cord, had completely disappeared.

Warty Growths on the vocal cords removed with the forceps.

Case 9.—William J., æt. 40, a waiter, applied to me in May, 1863, on account of hoarseness of five years' standing. His general health was good, but fifteen years before, he had a primary venereal sore. He had never suffered from any secondary symptoms. The voice was harsh, but not suppressed, and with the laryngoscope, a large thin, flat, membranous growth was seen to project horizontally from each vocal cord, and to meet in the center. On account of the pendulous condition of the epiglottis, it was difficult to get an extensive view of the larynx, and consequently the growths could not be seen in their entirety. The appearance is shown in Fig. 24. The smallness of the laryngeal

FIG. 24.

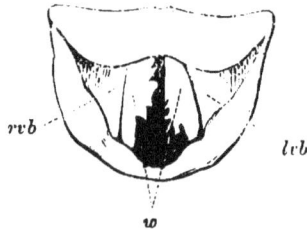

Fig. 24.—WARTS ON THE VOCAL CORDS.

(*This View of the Larynx was obtained by raising the Epiglottis with the Pincette.*)

rvb. Right ventricular band.
lvb. Left ventricular band.
w. Warts on the vocal cords.

aperture was still more inconvenient in operating, and it was only after several unsuccessful attempts that I managed to remove a small portion of the growth on the right vocal cord. Under these circumstances, I endeavored to divide the left growth through its base, with my laryngeal lancet. After the operation, the patient left me, but soon returned, spitting up considerable quantities of blood. On examination with the laryngoscope, the mucous membrane was seen to be covered with blood, but the exact source of the hemorrhage could not be ascertained. I applied a strong solution of perchloride of iron to the interior of the larynx, and directed the patient to suck ice. The hemorrhage, however, which continued for some time—to an extent that was really alarming,—was ultimately arrested by the patient gargling with, and swallowing, a saturated solution of tannin. The first mouthful of the tannin that was swallowed stopped the bleeding entirely. A day or two after the operation, a careful examination of the larynx was made both by Dr. George Johnson and myself, but we were neither of us able to ascertain the source of the hemorrhage. I have since removed several fragments by using the horizontal blades of my forceps, and the patient's voice is now clear ; he still complains, however, of a slight tickling in the throat.

Hoarseness of seven years' standing caused by a polypus attached just above the anterior insertion of the vocal cords. The polypus was removed with the forceps and the voice restored.

Case 10.—Morris B., æt. 41, shoemaker, and formerly singer, applied at the Dispensary for Diseases of the Throat, Aug. 20, 1863. He stated that he had been extremely hoarse for several years, but had never suffered from complete loss of voice. He had had primary syphilis when he was sixteen. A physician had recommended him to have his uvula removed, but the operation had not improved his voice. A laryngoscopic examination showed that there was a yellowish-pink growth, about the size of a small bean, just above the anterior insertion of the vocal cords. It was movable (and therefore probably pedunculated), but the base was hidden by the tumor, and therefore its exact origin could not be ascertained. When the glottis was closed, the growth rested on the extremities of both the cords; sometimes, however, lying more on the

FIG. 25.

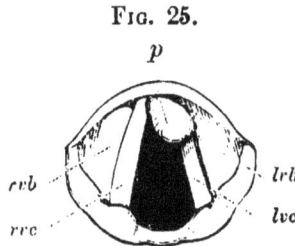

Fig. 25.—A SMALL POLYPUS, ATTACHED JUST ABOVE THE ANTERIOR INSERTION OF THE VOCAL CORDS.

> *rvb.* Right ventricular band.
> *rvc.* Right vocal cord.
> *lvb.* Left ventricular band.
> *lvc.* Left vocal cord.
> *p.* Polypus.

right, and sometimes on the left cord. The appearance is shown in Fig. 25.

Aug. 21.—I had the advantage of a consultation with Dr. George Johnson and Mr. Mason, who entirely concurred in my diagnosis.

Aug. 24.—In the presence of these two gentlemen, I removed the excrescence with my laryngeal forceps. The growth was successfully seized at the first trial, and all of it, except a small portion of its base, was brought away. After the operation, we examined the patient with a mirror, and the base of the growth covered with blood was indistinctly seen. I was disposed to remove this small remaining fragment, but after a consultation, it was thought better to leave it alone, under the idea that it would probably wither away.

Immediately after the operation, Dr. Johnson thought he noticed an improvement in the voice.

Aug. 26.--There being still a small portion of the base of the growth remaining, I removed it with the forceps. This case was completely cured, and at the end of a fortnight the man spoke perfectly well.

"The morbid growths," according to Dr. Andrew Clark, "consisted of three or four minute, shapeless pieces of yellowish color, streaked with red, and of a horny consistence. On account of their hardness, their structure could not be very easily determined. On the free surface, however, were several layers of thin, scaly epithelium, few of the elements of which exhibited any nuclei. In fact, but for the absence of cholesterine, the cell elements might have been most readily mistaken for those of cholesteatoma. Beneath the epithelial coverings, were minute extravasations of blood, and amorphous masses of a coagulated proteine compound." Though in this case the proteine compound had not developed fibers, the case was regarded by Dr. Clark as one of commencing "Fibro-Epithelial Growth."

Growths on both Vocal Cords (causing aphonia of nine months' standing), removed with the forceps.

Case 11.—Miss Mary B., æt. 30, was sent to me by Mr. Parsons, of Bridgewater, April 7, 1864. This patient lived in London, and after she had been suffering from loss of voice for some months, a distinguished physician recommended "change of air to her native place." On arriving there (Bridgewater) she was recommended to return back to London to see me, and the laryngoscope at once revealed the cause of the hoarseness.

A small growth was seen on the right vocal cord, and afterward, when the patient had been examined once or twice, another growth was perceived on the left cord, near to its anterior insertion. The appearance is shown in Fig. 26. The history of the case seemed to show that these

FIG. 26.

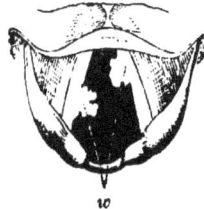

Fig. 26.—WARTS ON THE VOCAL CORDS.

The letter *w* points to the two growths.

growths originated in chronic laryngitis. After twenty attempts, only four of which were successful, the growths were entirely removed with the forceps. After the removal of the warts from the vocal cords, a small growth was seen lower down ; but as the voice was restored, no further treatment was adopted.

A small fish-bone removed with the forceps from the larynx of a child.

Case 12.—Caroline C., æt. 12, was brought to the Dispensary, February 8, 1864, by an elder sister. She was crying, and her sister stated that, on the previous evening, she had swallowed a herring-bone. She said that she felt the bone, whenever she swallowed. She had put her fingers down her throat to try and get it up, and had been sick several times. I examined her with the laryngoscope, but could only see that the mucous membrane of the larynx was much congested. Not being able to perceive the bone, I thought she had probably swallowed it, and recommended an inhalation of hot steam to relieve the irritation. I should mention that I passed a bougie into the stomach without difficulty. The next day the child was again brought to me. Her breathing was slightly stridulous, and a medical man, who had seen her in the mean time, had told her mother that "croup was coming on." On examining the patient with the laryngoscope, the ary-epiglottidean fold and ventricular band on the right side were seen to be much swollen, and of a bright red color, and a portion of the bone was distinctly seen lying across the right ary-epiglottidean fold, near the epiglottis. Apparent as it was, and easy as it seemed to seize, the greatest difficulty was experienced in getting it between the blades of the forceps. In the first attempt the mucous membrane was slightly wounded, and the bone became obscured by the blood. After an interval of half an hour, the bone again became visible, and it was fortunately grasped between the blades of the forceps. The bone was three-quarters of an inch long, and very thin. The patient complained of a little pricking in the throat for a day or two, but on the following Friday (the accident happened on Sunday) was quite well.

Contraction of the left Glosso-epiglottidean Fold from the healing of an ulcer; extreme dysphagia; relief from division of the structure.

Case 13.—Charlotte D., a married woman, æt. 24, was sent to me by Mr. Shillitoe, June 18, 1864. She stated that since November, 1863, she had not been able to swallow a particle of solid food; that she had lived entirely on liquids; and that bread soaked in milk was the nearest approach to a solid which she was able to get down. Even in swallowing liquids, a portion "went the wrong way," and she never attempted to drink, or rather sip, anything, without a violent and prolonged fit of coughing. Her symptoms all dated from an attack of ulcerated sore throat which she had had in October. Five years previously she had suffered from primary syphilis, and since then had had secondary symptoms.

On looking into the throat, numerous white cicatrixes were seen on the posterior wall of the pharynx, and on using the laryngoscope, the left side of the epiglottis was seen to be drawn upward, forward (toward the tongue), and slightly inward toward the median line. I at first thought that the left ary-epiglottidean fold was adherent to the fauces, but on subsequent examination, I found that such was not the case. The left glosso-epiglottidean fold was seen to be greatly thickened, white, prominent, and shortened, and it was obvious that this was the principal cause of the dysphagia. The dense band—raised not less than a quarter of an inch—could be felt with the finger. A bougie was passed down into the stomach without any difficulty, so that there was no doubt that the difficulty of swallowing depended mainly on the non-closure of the epiglottis over the larynx. I had the advantage of the opinion of Dr. Smyly, of Dublin, in this rare and difficult case. I at once determined on dividing the cicatrix, but its position in the antero-posterior diameter of the larynx made it difficult to use the lancet. With the laryngeal scissors, however, in two operations I succeeded in dividing it. There was very little hæmorrhage, and the patient, after a few weeks, was apparently able to swallow as well as most people; she still, however, complained of slight difficulty.

I have not seen this patient now for some months, but should not be surprised at her return any day, with her old symptoms. I should mention, that the epiglottis did not recover an entirely normal position; but instead of hanging very obliquely across the laryngeal aperture, the free edge of the valve became nearly horizontal. The appearance of the epiglottis, when the case first came under treatment, is shown in the annexed cut.

Fig. 27.

Fig. 27.—Contraction of the left Glosso-epiglottidean Fold, causing obliquity of the epiglottis, non-closure of the valve over the larynx, and (consequently) extreme dysphagia.

Since this case occurred, I have had the lancet modified, so that the cutting edge of the blade can be made to move in the antero-posterior or lateral diameter, according to the wish of the operator. Still, for dividing growths or strictures in the former direction, more power is obtained with the scissors.

CHAPTER VIII.

THE MANIPULATION OF LARYNGEAL INSTRUMENTS, AND SOME CONCLUDING REMARKS ON LARYNGO-SCOPY.

SECTION I.—*Method of employing Laryngeal Instruments.*

IN applying remedies to, or operating on the larynx, the practitioner should introduce the laryngeal mirror with his left hand, so that he may have his right hand free for using the necessary instrument. In employing the laryngeal brush, the operator should hold it like a pen, and introduce it quickly but steadily to the desired spot. In using any other instrument, the operator should hold it between the thumb and second finger, so that the index finger remains free, to press on the spring in the anterior and upper part of the handle.

Most of the laryngeal instruments invented by foreign physicians have the spring on which their action depends, situated at the posterior extremity

of the handle. The instruments are directed to be
held between the first and second fingers, while the
thumb pushes in the spring at the end of the handle.
In this method, the back of the operator's hand is
directed toward the patient's face, and half the
mouth is covered by the operator's third, fourth, and
fifth fingers; or the instrument may be held between
the first two fingers, and the spring at the posterior
part of the handle, may be made to act by the
pressure of the palm of the hand. A cross-bar near
the anterior extremity of the handle, on which the
two fingers can rest, facilitates the employment of instruments of this construction, but the great objection
to them is, that the pressure forward on the spring
at the end of the handle greatly alters the position
of the point of the instrument. In using instruments
made after my model,—that is to say, with the spring
on the upper and anterior part of the handle—the
position of the extremity introduced into the larynx
is not altered by the slight pressure of the index
finger on the spring. This is a matter of the
greatest importance in using the lancet or forceps.

Before dismissing this subject, I would call attention to the unity of action of all my laryngeal instruments. He who learns to use one, can use all; and the constant repetition of a particular act gives it a delicate precision, which is not otherwise attainable.

SECTION II.—*Concluding Remarks on Laryngoscopy.*

In this little work, many instruments and various kinds of apparatus have been described and recommended; but before concluding I would remark, that, with very few, and very simple appliances, the most satisfactory results may be accomplished,—not only in the diagnosis, but in the treatment of laryngeal disease. I have already observed that many of the most valuable laryngoscopic investigations have been made with a common moderator lamp, and I would call attention to the fact, that laryngeal growths have been removed with forceps of the most simple description. The forceps which Dr. Fauvel has several times used with success, has no complicated mechanism, and the instrument used in Dr. Russell's case was "an ordinary pair of curved forceps." Those who do not intend to take up the subject from a special point of view, but merely wish to use the laryngoscope in general practice, will do well to provide themselves with a reflector, a couple of laryngeal mirrors, a light-concentrator (which can be used with different kinds of lamps), a few laryngeal brushes, and my laryngeal galvanizer. A very large proportion of laryngeal diseases can be treated with the brush alone, and obstinate cases of functional aphonia cannot resist the internal application of galvanism. An additional recommendation to these instruments is, that even when employed injudiciously or ineffec tively, they are not likely to do any harm. It is only after the eye and the hand have had much practice in applying remedies to the larynx, that the lancet can be used with safety, or the forceps with effect. In conclusion, "I feel it a duty to remark," with Dr. Johnson, "upon the possibility that the larynx may get too much of local treatment. The laryngoscope has brought this organ so completely within our reach, that we are all exposed to the temptation of being too meddlesome. If we can avoid the error to which I have here alluded, the introduction of the laryngoscope will be an unmixed good both to ourselves and to our patients, and it will soon be acknowledged to be one of the most valuable additions that have ever been made to our means of diagnosis and treatment."

The following works bearing on the use of the Laryngoscope may be consulted:

Czermak, "Der Kehlkopfspiegel und seine Verwerthung für Physiologie und Medizin." Zweite Auflage. The first edition has been translated from a French version, and published by the New Sydenham Society, Vol. XI.—Türck, "Praktische Anleitung zur Laryngoskopie"—Wien, 1860; and "Recherches Cliniques sur diverses Maladies du Larynx"—Paris, 1862.—Semeleder, "Die Laryngoskopie und ihre Verwerthung für die Aerztliche Praxis." Wien, 1863.—Tobold, "Lehrbuch der Laryngoskopie." Berlin, 1863.—Bruns, "Die erste Ausrottung eines Polypen in der Kehlkopfshöhlen." Tübingen, 1863.—Sieveking, "Practical Remarks on Laryngeal Disease." Lond. 1862.—"Russell on Laryngeal Disease, &c." Lond. 1864.—Gibb, "Diseases of the Throat, &c." Lond. 1864.—"Walker on the Laryngoscope." Lond., T. Richards, 1864 (a most valuable pamphlet).—Dr. Johnson's "Lectures on the Laryngoscope," delivered at the College of Physicians. (These highly interesting lectures are shortly to be published.)—Reports of cases, and suggestions on the employment of the instrument, by Tonge, Mason, Ballard, and many others in the *Med. Times and Gaz.; British Med. Journ.;* and *Med. Circ.* Also some interesting reviews on the subject in the *Brit. and For. Med.-Chi. Rev.*, Oct., 1862, and July, 1864, and an original article in the same journal, Jan., 1863, by Mr. Winslow. To the latter I am indebted for several references bearing on the history of the invention of the Laryngoscope.

APPENDIX.

RHINOSCOPY.

THE idea of examining the posterior nares and Eustachian tubes, by placing a mirror at the back of the mouth, with its reflecting surface directed obliquely upward, appears to have occurred to Bozzini, Baumês, Wilde, and probably to others; but the practical application of this method of examination is undoubtedly due to Professor Czermak. The art of Rhinoscopy dates from a paper published by him in August, 1859.* Since then, Semeleder, Stoerk, Toltolini, Wagner, and others, have published various articles on the subject, but the first-named physician has especially developed, simplified, and proved the value and practicability of the art.

The principle of simple reflection alone is concerned in Rhinoscopy; but, as in the kindred art of

* *Wien Medizin Wochenschrift,* Aug. 6th, 1859.

Laryngoscopy, it is necessary to illuminate the parts desired to be inspected. A small mirror is placed at the back of the throat, at such an angle, that luminous rays falling on it, are reflected into the nares, while at the same time, the image formed on the mirror is seen by the observer.

For examining the posterior nares there are required, 1st, a small mirror; 2dly, a reflector; 3dly, a tongue-spatula; and 4thly, a palate-hook for raising the uvula, and pulling it forward. The rhinal mirror should be made like the laryngeal mirror, but its reflecting surface should not be more than five-eighths of an inch in diameter, and it should be fixed to its shank at a right angle. The reflector is the same as that used in Laryngoscopy. The tongue-spatula should have the part which is introduced into the mouth an inch longer, and should form a more acute angle with the handle than the ordinary instrument. The palate hook is generally made of German silver; it is about four inches long, narrow where it is fixed into the handle, and gradually getting broader at the opposite end, where it turns round at a right angle, and extends upward for a quarter of an inch. This instrument, though recommended by Czermak, is seldom of much use. He advises that it should be made of different sizes, bent at different angles, and in some cases fenestrated.

The examination should be conducted as follows: The lamp should have the same position as in Laryngoscopy, but the practitioner, in using the reflector, must throw the rays rather lower in the fauces. The patient should sit upright, with his head erect, or bent slightly forward (as suggested by Moura-Bourouillou), so that the uvula may hang forward; the patient is then directed to open his mouth widely, and the tongue pressed forward and downward with the spatula; the mirror is then introduced to the back of the throat (its upper border being a little below the uvula), so that the plane of the reflecting surface forms with the horizon an angle of about 130°. If the uvula is drawn upward and backward, the patient must be directed to expire gently or to produce some nasal sound. Straining and forced inspiration must be especially avoided. The practitioner will find it a good plan to introduce the small mirror between the anterior pillar and the uvula on

one side first, and then to withdraw it and introduce
it again in the same manner, on the opposite side.
In this way he will be able to inspect the whole of
the posterior nares, and by first slanting the mirror
a little toward one side, and then toward the other,
the orifices of the Eustachian tubes will become visible. After introducing the mirror in the way
described, the observer can steady it, by resting his
third and fourth fingers on the patient's lower jaw.
As already stated, Professor Czermak recommended
the use of the palate-hook for raising and drawing
forward the uvula. Where there is unusual insensi-
bility of the fauces, and the uvula is long, this operation may sometimes be accomplished, but I can
recall few instances where the procedure has facilitated my examination of the posterior nares. In
using the palate-hook, it must be first warmed, then
passed behind the uvula, and moved gently forward.
I have had an instrument made, in which the tongue-
spatula and mirror are combined, and I have since
ascertained that a similar rhinoscope is employed and
strongly recommended by Dr. Voltolini, of Breslau.
In the tongue-blade of the spatula, there is a narrow

groove in which the shank of the rhinal mirror runs, so that it can be passed a shorter or longer distance into the pharynx, according to the dimensions of the mouth. In cases where there is considerable space between the uvula and the posterior wall of the pharynx, and where the palate-spatula can be tolerated, this combination of tongue-depressor and mirror will be found useful.

The appearance seen with the rhinal mirror is very different to that which might be anticipated from an anatomical acquaintance with the osseous structures; and the parts, from their position, being seldom seen in the dissecting-room, and still more rarely in the dead-house, the novice in Rhinoscopy has but little preliminary knowledge, as regards the view of the nares from behind forward. It is seldom that the whole of the posterior nares can be seen with the mirror, as the soft palate generally eclipses the lower third. The wood-cut on the next page was taken from a drawing of the posterior nares of a woman, whose uvula had been removed some years previously.*

---

* The uvula has, however, been put into the drawing, in order to make it more intelligible.

The drawing is as accurate as possible, as regards form and size, but it is made up, out of a number of images, obtained by holding the mirror in different positions.

In the middle is seen the septum (*sn*). The mucous membrane in this situation is extremely thin, and appears almost white from the projection of the

FIG. 28.

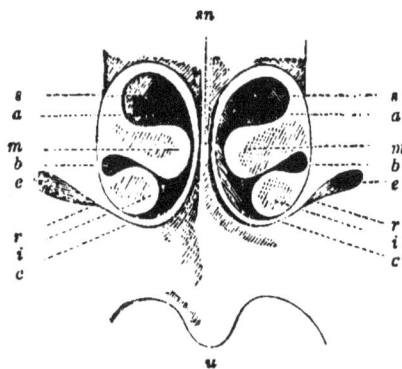

Fig. 28.—THE POSTERIOR NARES AS SEEN IN RHINOSCOPY.

sn. Septum nasi.
  s. Superior turbinated bone.
  m. Middle   "   "
  i. Inferior   "   "
  a. Superior meatus.
  b. Middle   "
  c. Inferior   "
  e. Eustachian orifice.
  r. Ridge between the Eustachian opening and lower
      border of the nasal fossa.

bone beneath it. The septum generally slants a little to one side, more often to the left, and it is rarely so symmetrical as in the annexed drawing.

Projecting from the outer wall of the nares on each side, and extending inward toward the septum, the middle turbinated bones (*m*), covered with pale mucous membrane (beneath which there are often a few muscular fibers), are seen, as two oblong tumors somewhat resembling polypi.* The superior ones are indistinctly seen, as narrow projections of somewhat triangular shape, the apex appearing to extend downward, inward, and backward. At the base of the nasal fossæ, the inferior turbinated bones are seen, as two pale, roundish, solid-looking tumors. They do not project quite so far toward the septum as the middle bones. The superior meati (*a*) are the largest of the three passages between the turbinated bones, the middle meati (*b*) are most distinctly seen toward the outer wall of the fossæ, and the inferior ones appear only as thin dark lines. On each side of the inferior turbinated bones, though further back and in a different plane, are the orifices of the Eusta-

* So much, indeed, do the turbinated bones resemble polypi, that not long since a surgeon—not much practiced in Rhinoscopy—brought me a case in which he imagined he had discovered a polypus in each nostril. It was only after demonstrating several other cases to him, that he became convinced that the supposed polypi were really the healthy turbinated bones.

11

chian tubes; they appear as two irregular openings looking downward and outward. "The upper and posterior edge of the trumpet-shaped opening of the Eustachian canal being beveled off, we see the inner surface of the anterior lip, buried in the pharyngeal wall, apparently of a lighter color than the surrounding mucous membrane, and having a yellow tinge communicated by the cartilage." (Walker.)

There is a prominent ridge (r) extending downward and inward from the lower edge of each Eustachian orifice (caused by the levator palati muscle on each side), and from the upper edge of the Eustachian opening is a depression in the mucous membrane, extending upward and inward. Beneath the nasal fossæ is the soft palate, uvula, &c.

The principal difficulty in Rhinoscopy is the length and breadth of the uvula, and too short a distance between the anterior pillars of the fauces and the posterior wall of the pharynx. The former obstacle may be overcome by following the directions already given, but the latter is insuperable. In a certain number of cases, it is quite impossible to practice Rhinoscopy, and it is generally easy by examining the fauces, and observing whether this space exists, to tell beforehand whether an inspection of the nasal fossæ is possible.

Though of comparatively limited, and rather difficult application, the art of Rhinoscopy proves useful in cases of obstruction of the nasal passages by polypi or thickened mucous membrane, in that most troublesome affection, ozœna, and in the various forms of ulceration of the hard and soft parts at the back of the nose. In cases of deafness dependent on obstruction of the Eustachian orifice, it not only enables the practitioner to diagnose the affection, but enables him to use the Eustachian catheter with safety and precision. 'There is a great difference of opinion, I believe, among aurists concerning the importance of treatment conducted through the Eustachian tube, but as regards the value of the mirror in carrying out this plan of treatment, there is no room for discussion. Drs. Gruber and Politzer, the well-known aural surgeons of Vienna, use the rhinal mirror in all cases of Eustachian disease. The following two cases are illustrative of the value of Rhinoscopy, and I would call attention to a very interesting case of nasal polypus published in the *Medical Circular*, January, 1864, in which Dr. Johnson used the rhinoscope with the greatest advantage to the patient.

Obstruction of the left nasal passage from thickening of the mucous membrane over the left middle turbinated bone. Constant sensation of stuffiness in the nose, and inability to blow the left nostril. Relief by local treatment.

Case 14.—Mrs. E., æt. 41, applied to me in June, 1863, on account of a constant feeling of stuffiness and troublesome pricking sensation in the nose. She said that for the

FIG. 29.

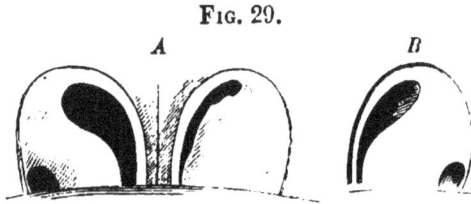

Fig. 29.—SWELLING OF THE MUCOUS MEMBRANE OF THE POSTERIOR NARES.

A. The mucous membrane over the middle turbinated bone, in the left nasal fossa (the right in the drawing), is seen to be so much swollen, that it nearly occludes the nasal passages.

B. The swelling of the mucous membrane has in part, but not entirely, subsided.

last three years she always felt as though she had a cold in her head, but there was no discharge. She had also a constant inclination to sniff through the left nostril. She had taken a great deal of medicine, and tried sniffing alum through the left nostril, but without any benefit. The physical conformation of parts was favorable to Rhinoscopy, and with the mirror the mucous membrane over the middle

turbinated bone on the left side was seen to be so much
swollen that it completely blocked up the left nasal passage. The mucous membrane was of a deep red color. With
a curved brush I applied various caustic solutions to the
affected part. Solutions of nitrate of silver, sulphate of
copper, and iodine were each used for some weeks, but the
greatest benefit resulted from the last-named agent. At
the end of four months, the subjective features of the disease had entirely given way, and the patient considered
herself cured. All the inflammation had subsided, but the
mucous membrane, though no longer at all congested, was
still a little swollen. Four months after the last application to the nares, the patient still felt perfectly well. The
appearance of parts when first seen, is shown in Fig. 29,
A, and the improvement after treatment in Fig. 29, B.

*Ozæna of two years' standing, caused by ulcers on the vomer
and right middle turbinated bone. Cured by local treatment.*

Case 15.—Hy. W., æt. 41, a shoemaker, applied at the
Dispensary for Diseases of the Throat, January, 1864, on

FIG. 30.

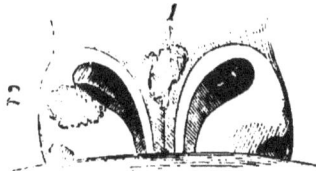

Fig. 30.—ULCERS IN THE POSTERIOR NARES CAUSING OZŒNA.

    1. Ulcer on the Septum.
    2. Ulcer on the right middle turbinated bone.

account of a constant discharge from the nose and throat,
from which he had been suffering for eleven months. The

larynx appeared to be healthy, but an ulcer was clearly seen at the upper part of the septum, and another on the right middle turbinated bone. On probing the ulcer on the septum with a piece of curved aluminium wire, the rough bony structure could be distinctly felt. On inquiry it was ascertained that two years previously the patient had suffered from constitutional syphilis.

The internal use of iodide of potassium and the local application of a solution of nitrate of silver effected a complete cure in six weeks.

These cases are selected from among others, which have been treated with equal advantage.

On the subject of Rhinoscopy, the reader may be referred to

Semeleder's interesting work, "Die Rhinoskopie und ihr Werth für die Aertzliche Praxis." Leipsig, 1862; Voltolini, "Eine monographische Arbeit zur fünzigjährigen Jubelfeier der Universität Breslau," Aug., 1861; and various papers by that author in "Virchow's Archiv," "Jahrbuch der Gesellschaft der Aerzte zu Wien;" "Deutsche Klinik," &c. (1860 and 1861).

INDEX.

www.ingramcontent.com/pod-product-compliance
Lightning Source LLC
Chambersburg PA
CBHW021811190326
41518CB00007B/551